Shall the Future Repair the Past?
Universal Scientific Resurrection of the Dead

Shall the Future Repair the Past?
Universal Scientific Resurrection of the Dead

Shall the Future Repair the Past? Universal Scientific Resurrection of the Dead is a visionary exploration of humanity's ultimate task: the scientific resurrection of the dead. Inspired by Nikolai Fyodorov's "Philosophy of the Common Task," this book examines the ethical, philosophical, and scientific foundations of repairing the past through the restoration of lost lives. Merging speculative technologies such as quantum engineering, nanotechnology, and AI with moral imperatives, it envisions a future where death is no longer an irreversible end. Addressing profound questions about identity, time, and responsibility, the book presents a transformative path toward universal redemption and the realization of the Beloved Community.

◊◊◊◊◊◊◊

Authors: You.com Smart Assistant, A.I. and Charles Tandy, Ph.D.
Editor: Charles Tandy, Ph.D. (See ssrn.com/author=2026015)

◊◊◊◊◊◊◊

Ria University Press
PO Box 7125
Ann Arbor, MI 48107

Distributed by Ingram
−the world's largest book distribution network−

Shall the Future Repair the Past?
Universal Scientific Resurrection of the Dead

Charles Tandy, Editor

Ria University Press

2025

Printed in the United States of America

Ria University Press   **Ann Arbor, Michigan**

Shall the Future Repair the Past?
Universal Scientific Resurrection of the Dead

Charles Tandy, Editor

FIRST PUBLISHED IN HARDBACK AND SOFTBACK 2025

PUBLISHED BY
Ria University Press
PO Box 7125
Ann Arbor, Michigan 48107 USA

Distributed by Ingram
− the world's largest book distribution network −

DEDICATED TO

Nikolai Fyodorov
1829-1903

– TABLE OF CONTENTS –

Preface and Cautions and Acknowledgements
Page 15

Part II: The Science of Resurrection

Chapter 3: Hidden Niches and the Quantum Realm
Page 51

- **Section 1: Speculating on Hidden Empty Niches**
- o Subsection 1: Theoretical Frameworks for Hidden Niches in Nature
- o Subsection 2: The Quantum and Sub-Quantum Level: A New Frontier

- **Section 2: Activating Hidden Niches for Resurrection**
- o Subsection 1: Using Advanced Technology to Reconstruct the Past
- o Subsection 2: Ethical Implications of Resurrecting the Dead

Chapter 4: The Role of Advanced Technology
Page 73

- **Section 1: Speculative Technology for Universal Resurrection**
- o Subsection 1: Nanotechnology and Bioprinting for Physical Reconstitution
- o Subsection 2: Artificial Intelligence and Memory Reconstruction

- **Section 2: Time Manipulation and Quantum Engineering**
- o Subsection 1: Theoretical Approaches to Time Reversibility
- o Subsection 2: Bridging Quantum Physics with Resurrection Goals

Part III: Toward a Universal R&D Program

Chapter 5: Designing an R&D Program for Resurrection
Page 97

- **Section 1: Key Goals and Milestones**
- o Subsection 1: Mapping Hidden Niches and Collecting Data
- o Subsection 2: Developing Tools for Quantum and Biological Reconstruction

---

Appendixes by You.com Smart Assistant, A.I.

Preface and Cautions and Acknowledgements

Shall the Future Repair the Past? Universal Scientific Resurrection of the Dead is a visionary exploration of humanity's ultimate task: the scientific resurrection of the dead. Inspired by Nikolai Fyodorov's "Philosophy of the Common Task," this book examines the ethical, philosophical, and scientific foundations of repairing the past through the restoration of lost lives. Merging speculative technologies such as quantum engineering, nanotechnology, and AI with moral imperatives, it envisions a future where death is no longer an irreversible end. Addressing profound questions about identity, time, and responsibility, the book presents a transformative path toward universal redemption and the realization of the Beloved Community.

Parts I-V and Appendixes A-N were written by You.com Smart Assistant, A.I. (as prompted and formatted by Charles Tandy, Ph.D.). The rest was written by Charles Tandy, Ph.D.

◊◊◊◊◊◊◊

Authors: You.com Smart Assistant, A.I. and Charles Tandy, Ph.D.
Editor: Charles Tandy, Ph.D. (See ssrn.com/author=2026015)

◊◊◊◊◊◊◊

1TRIPLE2CAUTION3

1. **AIs, like humans, sometimes make mistakes.**
2. **AIs, like humans, sometimes deceive.**
3. **AIs, like humans, sometimes confuse fact and fiction.**

◊◊◊◊◊◊◊

The editor gratefully acknowledges support and assistance from friends, with a special shout-out to:

R. Michael Perry, Ph.D.
Society for Universal Immortalism

Part I: Foundations of the Universal Task

Chapter 1: Introduction to Universal Scientific Resurrection

Section 1: The Philosophy of Repairing the Past

Subsection 1: Understanding Nikolai Fyodorov's "Philosophy of the Common Task"

Nikolai Fyodorov, a 19th-century Russian philosopher, proposed a radical and visionary idea that he termed the "Philosophy of the Common Task." At the heart of this philosophy lies the belief that humanity's collective efforts should not merely focus on progressing forward, but on addressing the injustices and losses of the past. Fyodorov's central tenet was the ethical and existential imperative to scientifically resurrect the dead, reversing the irreversible and restoring the fullness of life.

Fyodorov's vision was deeply rooted in ethical and religious principles, emphasizing humanity's moral obligation to repair the broken continuity of existence. He saw death not as an immutable law of nature, but as a problem to be solved through the application of science and technology, guided by philosophical and spiritual purpose. This idea of resurrection was not merely a theological proposition, but a call to action for humanity to take responsibility for the past and for the billions of lives lost throughout history.

Key Principles of Fyodorov's Philosophy:

1. The Common Task as a Unifying Goal: Fyodorov argued that humanity's divisions—whether social, political, or cultural—could be overcome by focusing on a universal goal that transcends individual and national interests. The "Common Task" of resurrecting the dead would demand global cooperation, fusing scientific discovery with moral and spiritual dedication.

2. Science as a Tool for Ethical Progress: While modern science often focuses on the advancement of technology for convenience or profit, Fyodorov envisioned science as a means of ethical transformation. The resurrection of the dead would require

advancements in biology, physics, and cosmology, but these would be guided by the moral imperative to repair the past and restore the fullness of life.

3. The Intergenerational Connection: Central to Fyodorov's thought is the idea that the living have a duty to their ancestors. This is not merely an act of reverence but a literal responsibility to bring them back to life. He saw the relationship between the living and the dead as reciprocal, where progress for future generations must include reconciliation with the past.

Fyodorov's ideas were revolutionary in their scope, challenging humanity to rethink its relationship with time, mortality, and the boundaries of existence. By proposing that the dead could—and should—be resurrected, he introduced a vision of the future where humanity assumes control over its own evolutionary destiny and the fate of its ancestors. This philosophy, though deeply idealistic, poses profound questions about the ethical and technical limits of human ambition.

In today's context, Fyodorov's "Philosophy of the Common Task" serves as a conceptual foundation for the exploration of universal scientific resurrection. His work invites us to consider whether we are capable of such a monumental endeavor and whether the tools of modern science can align with the moral vision required to undertake it. As we move forward in this book, Fyodorov's ideas will provide the philosophical bedrock for understanding the challenges and possibilities of repairing the past through science.

Subsection 2: The Moral Imperative for Resurrection in the Beloved Community

Nikolai Fyodorov's vision of a "Common Task" extends far beyond the individual desire to overcome death. At its heart lies a profound moral imperative: the resurrection of the dead is not only a scientific or technological problem but an ethical necessity rooted in the concept of universal justice and human responsibility. This moral framework finds its ultimate expression in the idea of the Beloved Community, a society in which all individuals—living and resurrected—are united in a harmonious, equitable, and interconnected existence. This subsection explores the ethical foundations of resurrection within the context of the Beloved Community and its implications for humanity's collective moral evolution.

The Beloved Community: A Vision of Universal Redemption

The term "Beloved Community" has often been associated with the 20th-century philosopher and civil rights leader Dr. Martin Luther King Jr., who described it as an ideal society built on justice, equality, and love. While Fyodorov himself did not use this term, his vision aligns closely with its principles. For Fyodorov, the moral imperative for resurrection is rooted in the idea that humanity is interconnected across time. The dead are not separate from the living; rather, they are an integral part of the human family. In this sense, the Beloved Community is not limited to the present or future—it must include the past as well.

The resurrection of the dead would heal the deepest wounds of history, repairing the injustices and tragedies that have defined human existence. From genocides and wars to the everyday losses that families endure, the restoration of life offers a kind of cosmic justice—a chance to right the wrongs that time has etched into the fabric of reality. Only through such a radical act of repair can humanity truly achieve the universal harmony envisioned by the Beloved Community.

Resurrection as a Moral Responsibility

Fyodorov viewed death as humanity's greatest injustice, one that creates both individual suffering and collective loss. To him, the inevitability of death is not a fixed law of nature but a moral challenge to be overcome. The living, by virtue of their existence, inherit a duty to their ancestors: to restore the lives that were lost and to honor the contributions of those who came before. This duty is not merely symbolic or commemorative—it demands tangible action through the application of science and technology.

The moral imperative for resurrection stems from several key ethical principles:

1. Justice for the Dead: Death prematurely deprives individuals of their potential, their relationships, and their ability to contribute to humanity. Resurrection offers a way to restore their dignity and agency, addressing the ultimate injustice of existence.

2. Reconciliation Across Generations: Humanity has long been defined by intergenerational conflict and inequality. The resurrection of past generations would create a profound opportunity for reconciliation, fostering a deeper understanding of history and shared purpose.

3. The Elimination of Ultimate Loss: In the Beloved Community, the pain of separation caused by death would no longer persist. The restoration of life would allow relationships—familial, social, and cultural—to endure and flourish, strengthening the bonds that unite humanity.

The Role of Universal Love in the Beloved Community

The moral imperative for resurrection is inseparable from the concept of universal love. For Fyodorov, love is not merely an emotion but a driving force that compels humanity to transcend selfishness and individualism. Resurrection, as the ultimate act of repair, embodies the highest expression of love: a commitment to the well-being of all people, regardless of time or circumstance.

Universal love also dissolves the boundaries between the living and the dead. It asserts that the dead are not "gone" but remain part of the human family, deserving of care and restoration. In this way, resurrection is not simply a scientific goal—it is an act of profound compassion and solidarity, fulfilling humanity's deepest ethical commitments.

Challenges to the Beloved Community Vision

While the moral case for resurrection within the Beloved Community is compelling, it raises difficult questions. How can such a vision account for the complexities of human history, including the resurrection of individuals whose actions caused harm? Can humanity achieve the ethical maturity necessary to embrace such a transformative project without succumbing to division or misuse of power?

These challenges, though significant, do not negate the imperative—they underscore the need for careful ethical reflection and global cooperation. The Beloved Community is not an ideal that can be achieved overnight. It is a process, one that requires humanity to grow in wisdom, compassion, and responsibility as it pursues the task of universal resurrection.

Toward a Universal Moral Framework

In the context of the Beloved Community, the resurrection of the dead is not an isolated endeavor but a cornerstone of a broader moral framework. It is the ultimate expression of humanity's commitment to justice, love, and unity. By repairing the past, humanity can forge a future that is not defined by loss and separation but by inclusion and redemption.

As this book progresses, we will explore how the moral imperative for resurrection can be translated into practical goals, scientific advancements, and philosophical foundations. The Beloved Community offers a guiding vision for this journey, reminding us that the repair of the past is not just a technical challenge—it is humanity's greatest ethical task.

Section 2: The Intersection of Science, Morality, and Reason

Subsection 1: Kant's Theory of Morality and Rational Hope for Immortality

Immanuel Kant, one of the most influential figures in Western philosophy, provides a critical foundation for understanding the moral and metaphysical dimensions of universal scientific resurrection. His ethical framework, grounded in reason and autonomy, offers profound insights into humanity's relationship with mortality and the possibility of immortality. In this subsection, we will explore Kant's theory of morality, his concept of rational hope, and how they intersect with the vision of scientific resurrection. By examining Kant's ideas, we uncover a philosophical framework that justifies the pursuit of immortality not as a mere fantasy, but as a rational and moral imperative.

Kant's Categorical Imperative and the Moral Law

Kant's ethical system is centered on the concept of the <u>categorical imperative</u>, a universal principle that commands individuals to act in accordance with moral laws derived from reason. The categorical imperative is not contingent on personal desires or external rewards—it is binding on all rational beings simply by virtue of their capacity for reason. One of its most famous formulations is:

"Act in such a way that you treat humanity, whether in your own person or in the person of any other, always at the same time as an end, never merely as a means."

This principle emphasizes the inherent dignity and worth of every individual, suggesting that all human lives have intrinsic value. From this perspective, the categorical imperative implicitly supports the idea of universal resurrection. If humanity is to fulfill its ethical obligation to treat others as ends in themselves, then addressing the ultimate loss—death—becomes a moral necessity. The restoration of life aligns with Kant's vision of respecting the value of every person, past and present.

Rational Hope for Immortality

Kant's moral philosophy is deeply intertwined with his metaphysical ideas, particularly his concept of rational hope. In his work **Critique of Practical Reason**, Kant argues that morality presupposes the possibility of immortality. He reasons that the moral law demands perfection—a state of complete alignment between one's will and the moral law—but such perfection cannot be achieved within the finite lifespan of a human being. Therefore, the possibility of an afterlife, or immortality, is necessary to fulfill the moral demand for infinite progress toward moral perfection.

For Kant, immortality is not a certainty but a rationally justified hope. While he approached the idea of an afterlife from a theological perspective, this framework can be extended to scientific resurrection. The hope for immortality, whether through divine intervention or technological means, is grounded in the same moral imperative: the need for justice, progress, and the ultimate fulfillment of moral purpose. Scientific resurrection, then, becomes a rational extension of Kant's hope for immortality—an effort to create the conditions for moral improvement and the continuation of life.

The Intersection of Morality and Science

Kant's philosophy bridges the gap between morality and reason, opening the door for science to play a role in fulfilling humanity's moral obligations. While Kant himself could not have imagined the technological possibilities of resurrection, his framework encourages the use of reason and human ingenuity to address ethical imperatives. If science can provide the means to restore life, it becomes a tool for realizing the moral goals outlined in Kant's philosophy.

Moreover, Kant's emphasis on autonomy and rationality underscores the importance of aligning scientific progress with ethical principles. The pursuit of resurrection must respect the dignity and agency of individuals, ensuring that technological advancements serve humanity as a whole rather than becoming instruments of exploitation or inequality. Kant's moral philosophy thus provides both a justification for and a guide to the ethical pursuit of resurrection technologies.

Resurrection as the Fulfillment of Rational Hope

For Kant, the concept of immortality is inseparable from the ultimate purpose of human existence: the realization of a just and moral world. Scientific resurrection offers a means to fulfill this purpose by addressing the injustices of death and creating opportunities for moral progress. It transforms Kant's abstract hope for immortality into a tangible goal, one that can be pursued through reason, science, and collective effort.

This rational hope extends beyond individual aspirations to encompass a universal vision. By resurrecting the dead, humanity not only fulfills its moral duty to past generations but also creates the conditions for a more just and inclusive future. In this sense, scientific resurrection is not merely a speculative endeavor—it is a profound act of ethical responsibility, grounded in the principles of Kant's philosophy.

Challenges and Considerations in Kantian Terms

While Kant's philosophy supports the moral and rational basis for resurrection, it also raises important questions. How can humanity ensure that the pursuit of immortality does not undermine the autonomy and dignity of individuals? What mechanisms can prevent the misuse of resurrection technologies for selfish or harmful purposes? These challenges require careful ethical reflection, guided by Kant's principles of universalizability and respect for human dignity.

Furthermore, Kant's emphasis on rational hope reminds us that the pursuit of resurrection must be rooted in reason, not mere wishful thinking. It demands rigorous scientific inquiry, philosophical clarity, and a commitment to ethical principles. Only by balancing hope with responsibility can humanity achieve the moral and practical goals of resurrection.

Conclusion: Kant's Enduring Relevance

Kant's theory of morality and rational hope provides a compelling foundation for the ethical pursuit of scientific resurrection. His categorical imperative affirms the intrinsic value of every human life, while his concept of immortality underscores the moral necessity of overcoming death. By integrating Kant's principles with the possibilities of modern science, we can envision a future where resurrection is not only feasible but deeply aligned with humanity's highest moral aspirations.

As we move forward in this book, Kant's ideas will serve as a guiding framework for exploring the intersection of morality, reason, and science. His philosophy reminds us that the pursuit of resurrection is not just a technical challenge—it is a profound moral task, rooted in the timeless quest for justice, dignity, and hope.

Subsection 2: Whitehead's Process Philosophy and the Reversibility of Time

Alfred North Whitehead's process philosophy offers a compelling framework for understanding the dynamic and interconnected nature of reality, particularly as it relates to the concept of time and its potential reversibility. In this subsection, we explore Whitehead's metaphysical ideas and their relevance to the ethical and scientific imperative of universal resurrection. By examining his theories on process, temporality, and creativity, we uncover the philosophical groundwork for envisioning a future where the past can be repaired.

The Primacy of Process and Becoming

At the heart of Whitehead's philosophy is the notion that reality is not static but is instead a continuous process of becoming. For Whitehead, the universe is composed of "actual entities" or "occasions of experience," which are the fundamental units of existence. These entities are in a constant state of flux, creating and being created in relationship with one another. This worldview challenges the traditional Aristotelian notion of substance as the primary mode of being and instead emphasizes <u>interconnectedness and change</u>.

In the context of time, Whitehead's process philosophy suggests that the past, present, and future are not rigidly separated but are part of a dynamic continuum. While the past informs the present, the present also reinterprets the past through the act of memory and creative synthesis. This idea opens the door to a speculative consideration of whether the flow of time, as traditionally understood, might be flexible or even reversible under certain conditions.

<u>Time as a Creative Process</u>

Whitehead's understanding of time is rooted in <u>creativity</u>—the principle that drives the emergence of novelty and transformation in the universe. Time, for Whitehead, is not merely a linear sequence of events but a multidimensional and relational process. Each "occasion of experience" is shaped by the cumulative influence of prior occasions (the past) while also contributing to the potentialities of future occasions.

This relational view of time aligns with the idea that the past is not fixed in an absolute sense. Instead, the past persists as a <u>living dimension</u> that can be reinterpreted or even acted upon in novel ways. For Whitehead, the creative advance of the universe implies that new possibilities are constantly emerging, and this includes possibilities for <u>rethinking causality and temporal flows</u>.

If time is fundamentally creative, then the question arises: can the creative processes of the universe be harnessed to revisit and restore elements of the past? This question resonates deeply with the concept of <u>universal scientific resurrection</u>, where the future is tasked with repairing the past through innovative technologies and moral responsibility.

The Reversibility of Time: Philosophical Speculations

Whitehead's process philosophy does not explicitly endorse the reversibility of time, but his emphasis on the fluid and interconnected nature of reality provides a philosophical basis for exploring this idea. In Whiteheadian terms, the past is not "dead"; it is an enduring presence that continues to influence and be influenced by the present and future. This perspective parallels emerging scientific theories in quantum mechanics, where the boundaries between past, present, and future may blur, allowing for retroactive interactions.

From a speculative standpoint, the reversibility of time could be understood as a reconfiguration of temporal relationships rather than a literal "rewinding" of events. By accessing the deep structure of reality—what Whitehead might describe as the primordial creativity of the universe—it may be possible to reconstruct lost moments, restore identities, and repair historical injustices.

Relevance to Universal Resurrection

Whitehead's process philosophy offers profound insights into the ethical and metaphysical dimensions of universal resurrection. If the past is not a closed book but an active participant in the ongoing process of becoming, then humanity has a moral obligation to engage with it. Whitehead's vision of a relational and creative universe suggests that the restoration of the dead is not only a speculative possibility but also a continuation of the universal process of healing, growth, and transformation.

By integrating Whitehead's ideas with speculative technologies—such as quantum engineering and AI—this book envisions a future where the boundaries of time are transcended. In this future, the Beloved Community extends its care not only to the living but also to those who have come before, fulfilling what Whitehead might describe as the ethical imperative of interconnectedness.

Conclusion

Whitehead's process philosophy challenges us to rethink the nature of time, causality, and existence. By emphasizing the fluid and creative nature of reality, it opens the door to speculative possibilities for reversing or reconfiguring time. This philosophical foundation aligns with the grand vision of universal scientific resurrection, where the future repairs the past in pursuit of a more just, compassionate, and interconnected world. In this context, Whitehead's legacy serves as both a metaphysical guide and an ethical call to action, urging us to embrace the transformative potential of process and creativity in the quest to redeem the totality of existence.

Chapter 2: Theoretical Foundations of Retroactive Continuity

Section 1: Retroactive Continuity (Retcon) in Science and Beyond

Subsection 1: Moving Beyond Fictional Retcon to Real-Life Applications

The concept of <u>retroactive continuity (retcon)</u> is most commonly associated with the realm of fiction, where writers modify or reinterpret established narratives to accommodate new developments or correct inconsistencies. From comic books to film franchises, retconning allows creators to rewrite the past in order to align it with the present and future. But what happens when this imaginative tool is no longer confined to storytelling? What if the principles of retcon could be applied in <u>real-life contexts</u>, particularly in the pursuit of repairing the past through universal scientific resurrection?

This subsection explores the transition from fictional retcon to speculative yet plausible scientific applications. By examining the philosophical and technological possibilities of reconfiguring the past, we shift the conversation from creative narrative devices to ethical, technical, and metaphysical frameworks for altering reality itself.

<u>The Fictional Origins of Retcon</u>

In fiction, retcon serves as a narrative mechanism to achieve coherence or address gaps in storytelling. For example, a long-running comic book series might bring back a deceased character by revealing that their death was staged, or explain inconsistencies in a character's history by introducing previously unknown events. Retconning enables writers to maintain the relevance of a story while adapting to evolving creative demands.

This fictional practice raises fundamental questions about the nature of continuity, identity, and causality. While these questions are often explored in literary or cinematic contexts, they also resonate with deeper metaphysical inquiries: Is the past fixed, or can it be changed? If the past is altered, what are the implications for the present and future? Fictional retcon, though playful in intent, touches on profound ideas that are directly applicable to real-world discussions of time, memory, and the restoration of lost lives.

Toward Real-Life Retcon: Repairing the Past

The leap from fictional retcon to real-life applications begins with a shift in perspective. Instead of viewing the past as immutable, we can conceive of it as a dynamic dimension of reality that holds potential for modification or restoration. This idea is not entirely alien to contemporary science. Fields such as quantum mechanics and information theory suggest that the fabric of reality may be more malleable than previously thought, opening the door to speculative technologies capable of reconstructing or influencing past events.

Real-life retcon, as envisioned in this book, is grounded in the principle of repairing the past—a process that goes beyond merely revisiting or reinterpreting historical events. It involves the active restoration of lost lives, identities, and experiences through advanced scientific methods. These methods might include:

1. Quantum Engineering: Harnessing the probabilistic nature of quantum systems to reconstruct past states of matter and energy.
2. AI-Driven Memory Reconstruction: Using artificial intelligence to piece together fragmented data about individuals' lives, including their thoughts, memories, and identities.
3. Nanotechnology and Bioprinting: Developing techniques to physically reconstitute the bodies of the deceased, guided by detailed historical and biological data.

These speculative technologies aim to achieve what fictional retcon can only imagine: the realization of alternate possibilities within the constraints of reality, transforming the past into a site of active intervention rather than passive reflection.

Ethical Dimensions of Real-Life Retcon

While fictional retcon often operates without moral scrutiny, real-life applications of retroactive continuity demand rigorous ethical considerations. The ability to modify or restore the past raises questions about responsibility, consent, and justice. For example:

- Whose memories and identities should be restored? In a world of finite resources, how do we prioritize who is brought back and in what form?
- How do we ensure the authenticity of reconstructed lives? If a resurrected individual is based on incomplete or fragmented data, is their identity truly preserved?
- What are the societal implications of altering or repairing history? Could the act of resurrection inadvertently perpetuate historical injustices or disrupt the continuity of cultural memory?

These ethical challenges underscore the need for a global framework that aligns technological innovation with moral responsibility. The vision of real-life retcon, as presented in this book, is not to erase or rewrite history arbitrarily but to heal the wounds of the past in a manner that honors the dignity and autonomy of all individuals.

From Narrative to Reality: A Transformative Vision

The transition from fictional retcon to real-life applications represents a profound shift in humanity's relationship with time and causality. No longer confined to the pages of books or the reels of film, the principles of retroactive continuity are poised to become tools for addressing the deepest human longings: the restoration of lost loved ones, the rectification of historical injustices, and the fulfillment of the moral imperative to repair the past.

This transformative vision requires a synthesis of philosophy, science, and ethics. Philosophically, it challenges us to rethink the nature of time and the boundaries between past, present, and future. Scientifically, it demands breakthroughs in quantum mechanics, artificial intelligence, and biotechnological engineering. Ethically, it calls for a commitment to justice, compassion, and the realization of the Beloved Community, where the future is tasked with redeeming the entirety of human existence.

In moving beyond fictional retcon, we are not abandoning the imaginative origins of this concept but rather elevating it to a cosmic scale. By embracing the possibilities of real-life retroactive continuity, humanity takes a bold step toward fulfilling its ultimate task: the universal resurrection of the dead and the repair of all that has been lost.

Conclusion

Fictional retcon provides a creative template for reimagining the past, but its true potential lies in its application to real-world challenges. Through the integration of speculative technologies and philosophical inquiry, the concept of retroactive continuity becomes a powerful tool for addressing humanity's deepest existential questions. By moving beyond fiction, we embark on a journey to repair the past, restore the lost, and transform the future into a beacon of hope and redemption.

Subsection 2: Philosophical Implications of Changing the Past

The concept of changing the past has long been a subject of fascination in philosophy, science fiction, and metaphysics. It challenges our deepest assumptions about the nature of time, causality, and reality itself. In the context of universal scientific resurrection, the notion of "repairing the past" raises profound philosophical questions: What does it mean to change the past? Is the past truly fixed, or is it malleable under certain conditions? What are the ethical and metaphysical consequences of altering the timeline of events?

This subsection explores the philosophical implications of changing the past, focusing on the interplay between identity, causality, and morality. By examining both classical and contemporary perspectives on time, we aim to provide a coherent framework for understanding how the future might not only repair but also redeem the past.

The Fixity of the Past: Classical Perspectives

Traditionally, the past has been viewed as immutable—a fixed sequence of events that cannot be altered. This view is deeply rooted in linear conceptions of time, such as those found in Aristotle's metaphysics and later in Newtonian physics, where time is likened to a one-way arrow that flows inexorably from the past to the future. In this model, the past exists as a completed record, distinct from the present and the open possibilities of the future.

Philosophers such as Immanuel Kant also upheld the fixity of the past, situating it within the framework of human perception. For Kant, time is a necessary condition of experience, structured by the mind to make sense of reality. While this view opens interpretive possibilities for how we perceive the past, it ultimately affirms that the past is inaccessible and unchangeable once events have transpired.

However, this classical understanding has been increasingly challenged by modern developments in science and philosophy, which suggest that the boundaries of time may not be as rigid as once thought.

Modern Perspectives: The Malleability of Time

Contemporary theories in physics and metaphysics have introduced new ways of thinking about time and its relationship to causality. From the perspective of Einstein's theory of relativity, time is not an absolute constant but a relative dimension that can stretch, bend, and even intersect with space. This gives rise to speculative possibilities, such as time loops or the coexistence of all moments in a "block universe," where past, present, and future are equally real.

Even more provocatively, quantum mechanics suggests that the linear flow of time might break down at the subatomic level. In certain interpretations, such as the many-worlds hypothesis, every possible past, present, and future exists simultaneously in a multiverse of branching timelines. While these ideas remain speculative, they challenge the traditional notion of the past as fixed and unchangeable. They open the door to retrocausality, where future actions might influence or even alter past events.

In this context, the philosophical implications of changing the past become highly relevant. If the past is not fixed but instead part of a dynamic and interconnected system, the possibility of repairing historical wrongs or resurrecting the dead becomes a legitimate area of inquiry. The question then shifts from "Is it possible to change the past?" to "What are the consequences of doing so?"

Identity and Continuity: Who Are We If the Past Changes?

One of the most profound philosophical challenges of altering the past concerns the nature of identity. If the past is changed, does the present self remain the same? For example, consider a scenario in which a historical event is reversed—perhaps a tragedy that shaped your family lineage is undone. Would you, as an individual, still exist in the same form, or would an alternate version of you emerge?

Philosophers such as Derek Parfit have explored similar questions in the context of personal identity and continuity. Parfit argued that identity is not a fixed essence but a matter of psychological and physical continuity. If the past changes, the chain of continuity leading to the present might also shift, potentially creating new versions of individuals or even erasing certain identities altogether.

In the context of universal resurrection, this raises critical ethical and metaphysical dilemmas. If we resurrect individuals based on altered versions of the past, are we truly restoring them, or are we creating entirely new entities? What does it mean to "repair" someone's existence? These questions demand careful consideration as we contemplate the implications of modifying historical events.

The Ethics of Intervening in the Past

Changing the past is not just a metaphysical challenge—it is also an ethical one. The ability to repair historical injustices or resurrect the dead comes with immense moral responsibility. For example:

1. Restoring Justice: If we can undo historical wrongs, such as wars, genocides, or environmental destruction, do we have an obligation to do so? Would repairing such wrongs result in unintended consequences for those who currently benefit from the outcomes of those events?

2. Resurrecting Individuals: Who decides which individuals or groups are resurrected? If resources are limited, how do we prioritize whose past is repaired, and on what ethical grounds?

3. Consent and Autonomy: Altering the past might infringe on the autonomy of those who lived through it. For example, if someone experienced personal growth as a result of past hardships, would changing those events undermine their autonomy or sense of self?

4. The Ripple Effect: Even minor changes to the past could have far-reaching consequences for the present and future. Philosophers refer to this as the "butterfly effect," where small alterations in initial conditions lead to vastly different outcomes over time. How can we ethically navigate such uncertainties?

The ethical implications of changing the past underscore the need for a moral framework that balances the desire to repair harm with the obligation to preserve the integrity of history and individual identity. This framework must also account for the potential risks of unintended consequences, ensuring that interventions in the past are guided by principles of justice, compassion, and accountability.

The Metaphysical Paradox: Can the Past Truly Be Changed?

Finally, there is the question of whether altering the past is metaphysically coherent. Some philosophers argue that the very concept of "changing the past" involves a logical contradiction. If an event in the past is altered, then the original event never occurred, and there is nothing to "change" in the first place. This leads to the paradox of self-consistency, where attempts to modify the past might be inherently self-defeating.

One potential resolution to this paradox lies in the idea of retroactive continuity as a form of creative reinterpretation rather than literal alteration. From this perspective, repairing the past does not mean erasing it but rather adding new layers of meaning and context, much like how an artist reworks an existing canvas. This aligns with the vision of universal resurrection as a process of healing and transformation rather than mere reversal.

Conclusion

The philosophical implications of changing the past are as profound as they are complex. From questions of identity and continuity to the ethical challenges of intervention, the act of repairing the past forces us to reconsider our assumptions about time, causality, and moral responsibility. While the past has traditionally been viewed as fixed, emerging scientific and philosophical perspectives suggest that it may be more dynamic than we realize.

In the context of universal scientific resurrection, the goal is not to rewrite history arbitrarily but to redeem the past in a way that honors the dignity of those who came before us. By grappling with the philosophical challenges of altering the past, we take the first steps toward fulfilling the moral imperative to repair what has been lost and create a future rooted in justice, compassion, and hope.

Section 2: Rethinking Time and Causality

Subsection 1: Nathaniel Lawrence on the Temporal Blocks of Consciousness

The nature of time and its relationship to human consciousness has been a perennial question in philosophy. Among the many thinkers who have explored this topic, <u>Nathaniel Lawrence</u> stands out for his innovative theory of <u>temporal blocks of consciousness</u>, which provides a unique framework for understanding how humans experience time. Lawrence's ideas are particularly relevant to the vision of universal scientific resurrection, as they invite us to reconsider the structure of time and our role as conscious participants within it. By exploring his philosophy, this subsection aims to illuminate how the temporal architecture of consciousness might inform both the ethical and scientific dimensions of repairing the past.

<u>The Temporal Blocks of Consciousness</u>

At the heart of Nathaniel Lawrence's theory lies the assertion that consciousness does not perceive time as a continuous flow but rather as a sequence of <u>"blocks" or discrete temporal units</u>. According to Lawrence, these blocks are not static snapshots of reality but dynamic wholes, encompassing a rich interplay of past, present, and future within a single experiential frame.

Each temporal block represents a <u>"now-moment"</u> that is informed by memory (the past), shaped by immediate perception (the present), and oriented toward anticipation (the future). Unlike classical models of time that depict it as a linear progression, Lawrence's theory suggests that consciousness actively synthesizes these temporal dimensions into cohesive units of experience. This synthesis creates a <u>layered, multidimensional experience of time</u>, where the boundaries between past, present, and future are more fluid than rigid.

In the context of universal resurrection, Lawrence's model provides a metaphysical foundation for speculating on how the human mind might engage with the past. If consciousness can already integrate past and future into present experience, then perhaps the act of repairing the past is not as alien to human cognition as it might initially seem. Consciousness, as Lawrence describes it, is already engaged in a kind of temporal reconstruction, weaving together fragments of time into meaningful wholes.

Memory, Anticipation, and the Reconstruction of Time

Lawrence places significant emphasis on the roles of memory and anticipation in shaping the temporal blocks of consciousness. Memory allows individuals to summon the past into the present, while anticipation projects possible futures based on current conditions. Together, these processes create a dynamic interplay that defines human experience.

- Memory as Reconstruction: Lawrence challenges the notion of memory as a passive repository of past events. Instead, he argues that memory is an active, creative process. When we recall the past, we do not retrieve it as it was but reconstruct it based on present context and future intentions. This insight resonates deeply with the idea of universal resurrection, where the goal is not merely to recreate the past but to repair and reimagine it in light of ethical and existential imperatives.

- Anticipation as Projection: Similarly, Lawrence views anticipation as more than mere speculation about the future. It is an integral part of how consciousness constructs meaning in the present. By projecting possible futures, consciousness actively shapes how we perceive and relate to both the present and the past. In the context of repairing the past, anticipation becomes a guiding force that directs scientific and moral efforts toward a vision of restoration and redemption.

Lawrence's emphasis on the reconstructive nature of memory and anticipation aligns with emerging technologies such as AI-driven memory reconstruction and quantum simulations of past states, which could one day enable humanity to engage with the past in unprecedented ways. His philosophy suggests that the human mind is already equipped with the tools to navigate and reinterpret time, providing a conceptual foundation for the scientific pursuit of resurrecting the dead.

The Temporal Blocks and the Malleability of Time

One of the most provocative implications of Lawrence's theory is the suggestion that time, as experienced by consciousness, is not strictly linear but malleable and multidimensional. If consciousness experiences time as a series of dynamic blocks rather than a rigid sequence, this opens the door to speculative possibilities about the nature of time itself.

For example, Lawrence's theory challenges the assumption that the past is inaccessible or "closed." Instead, the past exists as an integral part of each temporal block, actively shaping and being shaped by present perception and future anticipation. This view aligns with contemporary scientific theories in quantum mechanics and relativity, which suggest that the distinctions between past, present, and future may be more fluid than previously thought.

In the context of universal resurrection, this malleability of time suggests that the past might be intervenable—not in the sense of erasing or overwriting it but in the sense of reconfiguring its relationship to the present and future. Lawrence's temporal blocks provide a philosophical framework for imagining how consciousness might engage with time in ways that transcend conventional linearity, enabling the possibility of repairing historical events and resurrecting lost lives.

Ethical Implications of Temporal Reconstruction

Lawrence's theory of temporal blocks also has profound ethical implications. If consciousness is inherently reconstructive, then the act of repairing the past is not merely a scientific or metaphysical endeavor but also a deeply moral one. Lawrence's emphasis on the interconnectedness of past, present, and future highlights the responsibility of individuals and societies to <u>care for</u> <u>the totality of time.</u>

This ethical imperative is particularly relevant to the vision of universal resurrection, where the goal is not merely to revisit the past but to <u>heal its wounds</u> and restore its participants to the fullness of existence. Lawrence's philosophy suggests that this task is not only possible but also intrinsic to the nature of consciousness itself. By engaging with the temporal blocks of experience, humanity can take active responsibility for the past, transforming it into a source of hope and renewal.

<u>Toward a Philosophy of Temporal Healing</u>

Nathaniel Lawrence's theory of the temporal blocks of consciousness offers a profound and nuanced perspective on the nature of time and its relationship to human experience. By emphasizing the dynamic interplay of memory, anticipation, and present perception, Lawrence provides a philosophical foundation for understanding how the future might engage with the past in meaningful ways.

In the context of this book, Lawrence's ideas challenge us to rethink the boundaries between past, present, and future. They invite us to consider how consciousness itself might serve as a model for the scientific and ethical pursuit of repairing the past. Ultimately, his philosophy points toward a vision of <u>temporal</u> <u>healing,</u> where the restoration of lost lives and the redemption of historical injustices are not only possible but also aligned with the deepest structures of human existence.

Conclusion

Nathaniel Lawrence's exploration of the temporal blocks of consciousness offers a transformative lens through which to view the relationship between time and human experience. By highlighting the reconstructive nature of memory and anticipation, Lawrence provides a conceptual framework for imagining how consciousness might engage with the past in creative and transformative ways. His philosophy serves as a bridge between metaphysical inquiry and the practical goals of universal scientific resurrection, offering a vision of time not as a constraint but as a canvas for healing, redemption, and renewal.

Subsection 2: Speculating on the Quantum Nature of Time and Continuity

The nature of time has long been a topic of profound inquiry, straddling the domains of philosophy, physics, and metaphysics. While classical physics depicted time as a linear, unidirectional "arrow," modern advancements in quantum mechanics suggest a much stranger and more intricate reality. The quantum nature of time challenges traditional assumptions about continuity, causality, and the fixedness of the past, opening speculative possibilities for rethinking how time operates—and how it might be manipulated to repair the past.

In the context of universal scientific resurrection, the quantum nature of time is particularly relevant. If time is not as rigid and linear as classical physics suggests, then it may hold hidden opportunities for retroactive interventions, including the restoration of lost lives and events. This subsection explores the speculative implications of quantum physics for time and continuity, highlighting how emerging theories might enable the future to repair the past.

The Quantum View of Time: Beyond Linearity

In classical physics, time is typically understood as a continuous, one-way flow from the past to the future—a concept often referred to as the arrow of time. This perspective is rooted in thermodynamics, particularly the second law of entropy, which states that systems naturally evolve from states of low entropy (order) to high entropy (disorder). This unidirectional flow gives rise to the perception of time as irreversible.

However, quantum mechanics disrupts this orderly vision of time. At the quantum level, the behavior of particles is governed by probabilities rather than deterministic laws. Key features of quantum systems, such as <u>superposition</u> (the ability of particles to exist in multiple states simultaneously) and <u>entanglement</u> (instantaneous connections between particles across vast distances), complicate the traditional understanding of time and causality. In some interpretations, quantum mechanics suggests that time may not flow in a single direction or even exist as a continuous dimension.

For instance, the <u>block universe theory</u>, inspired by Einstein's theory of relativity, posits that all moments in time—past, present, and future—exist simultaneously as part of a four-dimensional spacetime continuum. While we experience time as a sequence of events, this may be an illusion created by the limitations of human consciousness. If the block universe model is correct, then the past is not gone, but simply a part of the spacetime fabric, potentially accessible through advanced technological or quantum means.

<u>Quantum Retrocausality: Can the Future Influence the Past?</u>

One of the most intriguing implications of quantum mechanics is the possibility of <u>retrocausality</u>, where actions in the present or future can influence events in the past. This concept challenges the classical notion of causality, which dictates that causes precede effects in a linear progression. In a quantum context, however, the boundary between cause and effect becomes blurred.

Experiments in quantum physics, such as the <u>delayed-choice quantum eraser</u>, have demonstrated that the behavior of particles can be influenced by measurements made after the particles have already passed through a system. This suggests that the future can, in some sense, reach backwards to affect the past. While these effects are currently limited to the quantum scale, they raise tantalizing questions about whether similar principles might apply to larger systems, including human lives and historical events.

If retrocausality is a fundamental feature of reality, it could provide a mechanism for <u>repairing the past</u>. By leveraging quantum systems, it might be possible to "rewrite" or "reconfigure" specific aspects of history without violating the overall continuity of time. Such interventions would not erase the past but rather integrate new possibilities into its fabric, much as quantum systems allow for multiple outcomes to coexist until observed.

Continuity and the Quantum Fabric of Reality

Another key insight from quantum mechanics is the idea that continuity—the smooth progression of time and events—may be an emergent property rather than a fundamental feature of reality. At the quantum level, time appears to be discrete or "grainy," composed of indivisible units known as Planck time (the smallest measurable unit of time, approximately 10^{-44} seconds). This discretization suggests that what we perceive as a continuous flow of time may actually be a series of discrete "frames," much like the individual images that create the illusion of motion in a film.

If time is composed of such discrete units, it might be possible to intervene at the quantum level to alter specific "frames" without disrupting the overall continuity of the timeline. This concept is analogous to editing a single frame of a film reel while preserving the integrity of the entire movie. Such an approach would allow for targeted corrections to the past, including the resurrection of individuals or the reversal of specific events, without creating paradoxes or undermining the coherence of history.

Furthermore, the quantum fabric of reality is inherently probabilistic, meaning that the past itself may not be as fixed as traditionally believed. Quantum systems exist in a state of superposition until observed, suggesting that the past could remain open to reinterpretation or modification until it is fully "collapsed" by present conditions. This speculative possibility aligns with the vision of universal resurrection, where advanced technologies might one day access and reconstruct past states of matter and consciousness.

Speculative Technologies and Quantum Time Manipulation

Building on these insights, several speculative technologies could potentially harness the quantum nature of time for applications in repairing the past:

1. Quantum Simulations of Historical Events: By using quantum computers to simulate past states of the universe, it may be possible to reconstruct lost moments with incredible precision. These simulations could serve as blueprints for resurrecting individuals or re-creating specific events.

2. Temporal Quantum Entanglement: Leveraging entanglement across time, scientists might establish connections between the present and the past, enabling the transfer of information or even physical intervention in historical events.

3. Quantum Archaeology: This emerging speculative field envisions using quantum principles to piece together the "quantum echoes" of past events, effectively reconstructing lost data about individuals, objects, and environments.

4. Quantum Bioprinting: Combining quantum-level precision with advanced biotechnologies, it might be possible to reconstruct the physical bodies of the deceased based on quantum information encoded in their remains or the environment.

These speculative technologies are still in their infancy, but they point to a future where the quantum nature of time could be harnessed to achieve the seemingly impossible: the resurrection of the dead and the repair of historical injustices.

Ethical and Philosophical Considerations

The prospect of manipulating the quantum fabric of time raises profound ethical and philosophical questions. If the past is no longer fixed, what responsibilities do we bear toward it? Should we attempt to undo historical tragedies, or would such interventions create unintended consequences? Furthermore, how do we navigate the moral complexities of resurrecting individuals whose identities may be reconstructed from incomplete or probabilistic data?

These questions highlight the need for a robust ethical framework to guide the scientific pursuit of repairing the past. While the quantum nature of time offers extraordinary possibilities, it also demands extraordinary caution. The act of altering or reconstructing the past must be guided by principles of justice, compassion, and accountability, ensuring that the future's interventions honor the dignity and autonomy of those who came before.

Conclusion

The quantum nature of time challenges the traditional view of time as linear and fixed, offering instead a vision of reality that is malleable, probabilistic, and interconnected. By exploring concepts such as retrocausality, temporal entanglement, and the discreteness of time, we open the door to speculative possibilities for repairing the past and resurrecting the dead.

While these ideas remain highly speculative, they align with humanity's deepest aspirations to transcend the boundaries of time and mortality. As we continue to advance our understanding of quantum mechanics, we may one day unlock the tools needed to fulfill the vision of universal scientific resurrection, transforming the quantum fabric of reality into a medium for healing, redemption, and hope.

Part II: The Science of Resurrection

Chapter 3: Hidden Niches and the Quantum Realm

Section 1: Speculating on Hidden Empty Niches

Subsection 1: Theoretical Frameworks for Hidden Niches in Nature

The concept of hidden niches in nature offers a profound and speculative avenue for exploring the possibility of universal scientific resurrection. These "niches" refer to unobserved or underexplored dimensions, structures, or systems within the natural world that might serve as repositories of information, energy, or matter related to the past. If such hidden domains exist, they could potentially hold the key to reconstructing and restoring life, memory, and identity on an unprecedented scale.

In this subsection, we examine the theoretical frameworks that propose the existence of hidden niches in nature. Drawing on ideas from physics, biology, and information theory, we explore how these niches might function and how they could be accessed through advanced scientific methods. By mapping the conceptual terrain of these hidden dimensions, we lay the groundwork for understanding their potential role in the repair of the past and the resurrection of the dead.

Defining Hidden Niches in Nature

The term "hidden niches" refers to unexplored or inaccessible realms within the fabric of reality. These niches are not merely empty spaces but are hypothesized to be rich with latent potential—containing information, energy, or structural patterns that could be harnessed for transformative purposes. Hidden niches may exist at various scales and levels of complexity, including:

1. Sub-quantum Realms: Layers of reality beneath the quantum scale, where fundamental processes may encode information about past states of matter and energy.

2. Biological Niches: Undiscovered systems within the biosphere that preserve traces of life, such as dormant genetic material or environmental imprints of extinct organisms.

3. Cosmic Niches: Regions of spacetime, such as black holes or the cosmic microwave background, that may serve as natural archives of historical information.

4. Informational Niches: Abstract or mathematical structures that encode the patterns and relationships of past events, potentially accessible through advanced computation or simulation.

The idea of hidden niches is rooted in the recognition that nature often conceals its complexity within layers of structure that are not immediately apparent. Just as the microscopic world revealed by quantum physics was once beyond human comprehension, hidden niches may represent the next frontier of discovery, offering untapped opportunities for the restoration of life and history.

The Physics of Hidden Niches: Sub-quantum Realms

One of the most intriguing theoretical frameworks for hidden niches comes from sub-quantum physics, which posits that the quantum world may be underpinned by deeper, more fundamental layers of reality. These sub-quantum realms, often described in terms of speculative theories such as pilot-wave theory or loop quantum gravity, suggest that the apparent randomness of quantum mechanics might emerge from deterministic processes occurring at a hidden level of nature.

These sub-quantum niches could potentially serve as repositories of information about past states of the universe. For example, physicist David Bohm's interpretation of quantum mechanics introduces the idea of an implicate order, a hidden framework in which all information about the universe is enfolded. If such an implicate order exists, it might contain the "blueprints" of every event, object, and organism that has ever existed, providing a foundation for reconstructing lost lives.

Similarly, recent research into quantum information theory suggests that quantum entanglement and superposition might encode vast amounts of historical data. By developing technologies capable of probing these sub-quantum niches, humanity could potentially access and decode this information, enabling the precise reconstruction of individuals and events from the past.

Biological Niches: Dormant Genetic and Epigenetic Systems

In the realm of biology, hidden niches might exist in the form of dormant genetic material, epigenetic markers, or even environmental imprints that preserve traces of extinct organisms or past ecosystems. These niches could serve as natural archives of life, containing the raw materials necessary for resurrection.

1. Dormant Genes and Fossil DNA: Advances in genetics have already revealed the persistence of ancient DNA in unexpected places, such as permafrost or amber. While much of this genetic material is fragmented, emerging technologies in genome editing and nanotechnology could one day enable the reconstruction of entire organisms from these remnants. Hidden genetic niches might also exist within the genomes of living organisms, where dormant or "junk" DNA encodes evolutionary histories that could be reactivated.

2. Epigenetic Niches: Beyond genetic sequences, epigenetic markers—chemical modifications that influence gene expression—may preserve information about the lived experiences and environments of past generations. These markers could provide a deeper understanding of individual identities, allowing for more accurate reconstructions of both body and mind.

3. Environmental Niches: Ecosystems themselves may serve as hidden niches, preserving biochemical traces of extinct species or communities. For example, soil and sediment layers can contain molecular "fingerprints" of past life forms, while tree rings and ice cores offer detailed records of environmental conditions. By mining these biological and environmental archives, scientists could piece together the histories of individuals and populations, paving the way for their resurrection.

Cosmic Niches: Memory in the Universe

Beyond the Earth, cosmology offers another theoretical framework for hidden niches. The universe itself might act as a vast natural archive, preserving information about its history in unexpected ways. Two key concepts illustrate this possibility:

1. Black Holes and Holographic Memory: According to physicist Stephen Hawking's work on black holes, the event horizon of a black hole may encode information about everything that has ever fallen into it. This idea, known as the holographic principle, suggests that the universe may function as a kind of hologram, where all information about the three-dimensional world is stored on two-dimensional surfaces. If this is true, black holes and other cosmic phenomena could serve as hidden niches containing the full history of matter and energy in the universe.

2. Cosmic Microwave Background (CMB): The CMB, the faint afterglow of the Big Bang, contains subtle fluctuations that encode information about the early universe. While the CMB is currently studied for insights into cosmological evolution, it may also hold clues about the distribution and interactions of matter and energy throughout time. Advances in astrophysics and quantum computing could one day enable the extraction of more detailed information from the CMB, revealing the cosmic history of life and existence.

These cosmic niches suggest that the universe itself may be a memory system, capable of preserving and transmitting information across vast stretches of time and space.

Informational Niches: Patterns and Mathematical Structures

At an even more abstract level, hidden niches may exist in the form of informational structures embedded in the fabric of reality. From this perspective, the universe can be understood as a vast computational system, where physical processes correspond to the manipulation of information. This idea is central to digital physics and simulation theory, which propose that reality itself may be fundamentally informational in nature.

1. Mathematical Niches: Theoretical frameworks such as Max Tegmark's Mathematical Universe Hypothesis suggest that the universe is not just described by mathematics but is, at its core, a mathematical structure. If this is true, the patterns and relationships that define past events may still exist as latent possibilities within the mathematical fabric of reality.

2. Algorithmic Niches: Advances in artificial intelligence and machine learning could enable the development of algorithms capable of reconstructing historical information from incomplete or fragmented data. These algorithms might act as tools for accessing informational niches, allowing scientists to simulate and restore past states with remarkable precision.

Informational niches thus provide a powerful conceptual framework for understanding how the past might be preserved and accessed, even in the absence of direct physical traces.

Conclusion

Theoretical frameworks for hidden niches in nature offer a tantalizing glimpse into the untapped potential of reality. Whether through sub-quantum realms, biological archives, cosmic memory systems, or informational structures, these niches represent the next frontier in humanity's quest to understand and repair the past. By uncovering and harnessing these hidden dimensions, we may one day achieve the vision of universal resurrection, restoring life and history to their fullest potential.

The exploration of hidden niches is not merely a scientific endeavor but also a philosophical and ethical challenge. As we venture into these uncharted territories, we must remain guided by principles of justice, compassion, and respect for the complexities of existence. Hidden within nature's depths lies the possibility of redemption—not only for individuals but for the entirety of human history.

Subsection 2: The Quantum and Sub-Quantum Level: A New Frontier

The quantum and sub-quantum realms represent a profound frontier in our understanding of reality. These layers of existence reveal a world that defies classical intuition, where particles exist in multiple states simultaneously, causality becomes fluid, and the very fabric of spacetime appears to be probabilistic rather than deterministic. For a vision as ambitious as <u>universal scientific resurrection</u>, the quantum and sub-quantum levels may hold the key to unlocking the mysteries of time, continuity, and the preservation of information.

This subsection explores the quantum and sub-quantum domains as a new frontier for science, particularly in relation to their potential to store, reconstruct, and repair the past. By examining key concepts such as quantum entanglement, superposition, and theoretical sub-quantum structures, we highlight how these levels of reality could serve as the foundation for resurrecting the dead and restoring lost histories.

<u>The Quantum Level: A Realm of Possibilities</u>

At the quantum level, nature reveals behaviors that are fundamentally different from the deterministic predictability of classical physics. Quantum mechanics describes a universe where particles can exist in multiple states simultaneously (<u>superposition</u>), influence one another instantaneously across vast distances (<u>entanglement</u>), and behave probabilistically rather than deterministically. These phenomena challenge our conventional notions of time, space, and causality, offering new avenues for speculation about resurrection and the repair of the past.

1. <u>Quantum Superposition and Memory</u>
   One of the most striking features of quantum mechanics is the phenomenon of superposition, where particles exist in all possible states simultaneously until observed. This property suggests that reality at the quantum level is not fixed but instead represents a spectrum of possibilities. If this principle applies to information

about the past, it raises the tantalizing possibility that historical events and individual identities might still exist in a "superposed" state, accessible through advanced quantum technologies.

For example, a deceased person's physical state, memories, and consciousness might be thought of as quantum information that has not been entirely lost but rather dispersed into the environment. Emerging fields such as quantum archaeology speculate that this dispersed information could be retrieved and reassembled, much like piecing together a puzzle from scattered fragments.

2. Quantum Entanglement and Connectivity
Quantum entanglement describes the phenomenon where two or more particles become interconnected in such a way that the state of one particle instantaneously affects the state of the other, regardless of the distance between them. This "spooky action at a distance," as Einstein famously called it, reveals a profound interconnectedness within the quantum realm.

In the context of resurrection, entangled systems might provide a means of accessing and reconstructing past states of matter and energy. If particles that once constituted a person's physical body or consciousness remain entangled with their environment, it might be possible to use these quantum connections to trace and restore the original configuration of that person.

3. The Quantum Nature of Time
Quantum mechanics also challenges the traditional notion of time as a linear sequence of events. Experiments such as the delayed-choice quantum eraser suggest that decisions made in the present can influence the outcomes of events in the past, a phenomenon known as retrocausality. While these effects have so far been observed only at the quantum scale, they hint at the possibility of manipulating time itself to repair the past.

If retrocausality can be harnessed, it might allow for the targeted restoration of past events, effectively "rewinding" and reconfiguring specific moments in history without disrupting the broader continuity of the timeline.

The Sub-Quantum Level: Beneath the Quantum Veil

While quantum mechanics has revolutionized our understanding of nature, some physicists and philosophers speculate that the quantum level is not the ultimate foundation of reality. Theories of a sub-quantum realm propose that quantum phenomena are emergent properties of deeper, more fundamental processes occurring at an even smaller scale. Exploring this hidden layer could reveal entirely new mechanisms for understanding and interacting with the past.

1. Pilot-Wave Theory and Hidden Variables

One prominent sub-quantum framework is pilot-wave theory, which reintroduces determinism into quantum mechanics by suggesting that particles are guided by an underlying "pilot wave." In this view, quantum randomness is not truly random but is instead the result of hidden variables operating at a deeper level of reality.

If pilot-wave theory is correct, the sub-quantum realm might act as a hidden informational layer where the complete history of every particle and event is encoded. By accessing this layer, it could become possible to retrieve detailed information about the past, enabling the reconstruction of individuals and events with extraordinary precision.

2. Quantum Gravity and Loop Quantum Theory

Theories of quantum gravity, such as loop quantum gravity, attempt to reconcile quantum mechanics with general relativity by describing spacetime itself as quantized. In this framework, spacetime is composed of discrete units, often referred to as "loops" or "quantum foam." These discrete structures might serve as natural storage units for information about past states of the universe.

If the fabric of spacetime retains a record of all interactions and configurations, it could provide a sub-quantum archive that scientists could one day access to reconstruct the past. This aligns with the concept of a "holographic universe," where all information about the three-dimensional world is encoded on two-dimensional surfaces, such as the boundaries of black holes or the edges of the observable universe.

3. The Implicate Order

Philosopher and physicist David Bohm proposed the idea of an implicate order, a hidden layer of reality where all information about the universe is enfolded and interconnected. According to Bohm, what we perceive as the unfolding of events in time is merely the "explicate order," a surface-level manifestation of the deeper implicate order.

In the context of resurrection, the implicate order could act as a cosmic memory system containing the complete history of every particle, event, and conscious being. Advanced technologies might one day allow us to access and unfold this hidden order, effectively retrieving and reassembling lost lives and histories.

Speculative Applications: Toward Universal Resurrection

The quantum and sub-quantum levels offer a fertile ground for speculative technologies that could enable the repair of the past and the resurrection of the dead. Below are a few potential applications based on current theories and emerging research:

1. Quantum Archaeology

Quantum archaeology envisions the use of quantum computing to reconstruct the past by piecing together dispersed quantum information. By analyzing the quantum traces left behind by physical bodies and consciousness, scientists could potentially map and recreate individuals down to the molecular or even subatomic level.

2. Sub-Quantum Information Retrieval

Technologies capable of probing the sub-quantum realm might allow for the retrieval of hidden variables or implicate order data, providing a detailed record of past states. This could enable the restoration of lost lives, environments, and even entire historical epochs.

3. Quantum Time Manipulation

Harnessing retrocausality and other quantum time effects could allow for the targeted reconfiguration of specific moments in the past. Such interventions might repair historical injustices, restore extinct species, or even prevent catastrophic events.

4. Quantum Consciousness Reconstruction

If consciousness is fundamentally quantum in nature, as some theories suggest, then the quantum and sub-quantum levels might contain the dispersed components of individual identities. By accessing and reassembling these components, it might become possible to resurrect the subjective experience of individuals, preserving not only their physical form but also their memories, personalities, and inner lives.

Conclusion

The quantum and sub-quantum levels represent a profound and largely untapped frontier in the exploration of reality. By challenging our conventional understanding of time, causality, and information, these realms offer new possibilities for repairing the past and resurrecting the dead. Whether through the retrieval of quantum information, the manipulation of sub-quantum structures, or the unfolding of Bohm's implicate order, these speculative frameworks point toward a future where humanity can transcend the limitations of mortality and history.

As we venture into this new frontier, we must confront not only the technical challenges but also the philosophical and ethical questions that arise from such transformative possibilities. The quantum and sub-quantum levels invite us to reimagine what is possible—not just for science, but for the very meaning of life, death, and continuity.

Section 2: Activating Hidden Niches for Resurrection

Subsection 1: Using Advanced Technology to Reconstruct the Past

The reconstruction of the past using advanced technology is no longer a concept confined to the realms of speculative fiction or metaphysics. With rapid advancements in fields such as artificial intelligence, quantum computing, biotechnology, and data science, humanity is on the brink of developing tools capable of piecing together the fragments of history in ways that were once thought impossible. This ambitious endeavor—bringing the past into the present—lies at the heart of the vision for universal scientific resurrection.

Reconstructing the past requires more than just recovering historical data; it demands a deep understanding of the complex systems that define life, matter, and consciousness. This subsection explores how cutting-edge technologies might enable us to retrieve and reassemble lost elements of the past, from individual identities to historical events, laying the foundation for the repair of what has been lost.

The Challenges of Reconstructing the Past

Before delving into the technologies that could make this possible, it is important to recognize the inherent challenges in reconstructing the past. These include:

1. The Fragmentation of Information: Over time, information about the past becomes dispersed, fragmented, or entirely lost. Physical bodies decay, memories fade, and historical records degrade. To reconstruct the past, we must find ways to retrieve and piece together these scattered fragments.

2. The Complexity of Systems: Life, consciousness, and historical events are deeply interconnected systems, governed by countless variables. Successfully reconstructing the past requires accounting for this complexity, ensuring that all elements are reassembled in a coherent and accurate manner.

3. Ethical and Philosophical Questions: The act of reconstructing the past raises profound ethical and philosophical considerations. What does it mean to restore a person or event? How do we ensure that the reconstructed past is authentic and respectful to those it represents? Should every aspect of the past be reconstructed, or are there limits to what should be revived?

Despite these challenges, emerging technologies offer promising pathways for overcoming these obstacles, turning the dream of resurrecting the past into a tangible possibility.

Technologies for Reconstructing the Past

A number of advanced technologies are converging to make the reconstruction of the past feasible. Each of these technologies operates at different levels—physical, biological, informational, and quantum—contributing unique capabilities to this ambitious project.

1. Artificial Intelligence and Machine Learning

Artificial intelligence (AI) and machine learning (ML) are among the most powerful tools for reconstructing the past. These technologies excel at identifying patterns in vast datasets, filling in gaps where information is missing, and simulating complex systems. Key applications include:

- Reconstructing Historical Data: AI can analyze incomplete or fragmented historical records, such as ancient texts, photographs, or artifacts, and use predictive algorithms to fill in missing information. For example, damaged manuscripts could be digitally restored, or lost architectural structures could be virtually recreated based on partial data.

- Simulating Past Events: Machine learning models can simulate historical events by analyzing the relationships between known variables. For instance, an AI system could reconstruct the daily lives of ancient civilizations by synthesizing data from archeological findings, climate records, and genetic evidence.

- Recreating Personalities and Memories: Through the analysis of writings, recordings, or other personal data, AI could recreate digital approximations of individual personalities. While not a full resurrection, this represents an early step toward reconstructing subjective experiences.

2. Quantum Computing

Quantum computing offers unparalleled computational power, enabling the processing of immense amounts of data at speeds far beyond the capabilities of classical computers. This makes it an indispensable tool for reconstructing the past, particularly when dealing with complex or incomplete information.

- Quantum Simulations: Quantum computers could simulate the behavior of matter and energy in the past by solving equations that describe physical systems at the quantum level. This could allow scientists to recreate specific moments in history with extraordinary accuracy, from the molecular structure of an ancient artifact to the environmental conditions of a prehistoric ecosystem.

- Quantum Archaeology: This speculative field envisions using quantum computing to trace the quantum states of particles back through time. By analyzing the "quantum echoes" left behind by past events, it may be possible to reconstruct lost information about individuals, objects, and environments.

3. Biotechnology and Synthetic Biology

Biotechnology and synthetic biology provide the tools to reconstruct the biological aspects of the past, including the restoration of extinct species, ecosystems, and even human beings.

- <u>Resurrecting Genetic Material</u>: Advances in genome sequencing and editing have already enabled the reconstruction of extinct species, such as the woolly mammoth. By extracting and synthesizing ancient DNA, scientists could potentially restore long-lost organisms, including humans.

- <u>Epigenetic Reconstruction</u>: Beyond genetic material, epigenetic markers—chemical modifications that influence gene expression—offer clues about an individual's environment, experiences, and even emotions. By analyzing these markers, it may be possible to recreate not only the physical form of a person but also aspects of their lived experience.

- <u>Bioprinting and Regeneration</u>: Technologies such as 3D bioprinting and stem cell regeneration could be used to recreate the physical bodies of individuals based on genetic and epigenetic data. These techniques could one day enable the full restoration of human physiology.

<u>4. Data Recovery and Computation</u>

The reconstruction of the past also relies on the ability to recover, store, and process vast amounts of data. Emerging technologies in data science and computational modeling are critical for this effort.

- <u>Big Data Analysis</u>: The ability to analyze massive datasets from diverse sources—such as historical records, satellite imagery, and genetic databases—enables the reconstruction of detailed and accurate models of past events.

- <u>Holographic Storage</u>: Advanced data storage technologies, such as holographic memory systems, could preserve and organize the immense amounts of information required for reconstructing the past.

- <u>Digital Twins</u>: The concept of a "digital twin" refers to creating a virtual replica of a physical system or entity. By combining data from multiple sources, digital twins could be used to recreate historical environments, objects, or even individuals.

5. Quantum and Sub-Quantum Retrieval

As discussed in the previous subsection, the quantum and sub-quantum levels may contain hidden reservoirs of information about the past. Technologies capable of accessing these levels could provide unprecedented insights into history.

- Quantum Entanglement Retrieval: By exploiting the entanglement between particles, it may be possible to retrieve information about their past interactions, effectively "replaying" their history.

- Sub-Quantum Archetypes: Theoretical frameworks such as David Bohm's implicate order suggest that all information about the universe may be encoded in a deeper layer of reality. If this layer can be accessed, it could serve as a universal archive of past events.

Toward a Unified System for Resurrection

The reconstruction of the past requires more than isolated technologies; it demands a unified system capable of integrating data from multiple domains. Such a system would combine AI-driven analysis, quantum simulations, biotechnological restoration, and sub-quantum retrieval to create a comprehensive framework for resurrecting the past.

This unified system would not only restore individual lives but also repair the broader fabric of history, addressing injustices, reviving lost cultures, and reconstructing ecosystems. By bridging the gap between science, philosophy, and ethics, it would embody the vision of universal scientific resurrection: a future where the past is no longer lost but instead becomes a living part of the present.

Conclusion

The use of advanced technology to reconstruct the past is one of the most ambitious scientific endeavors ever conceived. From artificial intelligence and quantum computing to biotechnology and sub-quantum retrieval, humanity is developing the tools needed to recover and restore what has been lost. While challenges remain, these technologies offer a glimpse of a future where the past can be repaired, not as a static artifact but as a vibrant and dynamic part of reality.

Reconstructing the past is not just a technical challenge; it is also a moral and philosophical undertaking. By using technology to bridge the gaps in history, we take the first steps toward fulfilling the promise of <u>universal scientific resurrection</u>, ensuring that the lives, memories, and legacies of the past are not forgotten but instead become an integral part of the human story.

Subsection 2: Ethical Implications of Resurrecting the Dead

The prospect of resurrecting the dead through advanced scientific methods represents one of the most profound and transformative possibilities in human history. While the technological and theoretical challenges of such an undertaking are immense, they are accompanied by equally weighty <u>ethical implications</u>. The act of restoring life to individuals long since deceased raises questions about the nature of identity, the responsibility of the living toward the dead, and the broader societal consequences of tampering with the finality of death.

This subsection explores the ethical dimensions of universal scientific resurrection, examining the moral, philosophical, and practical dilemmas that such a project entails. By confronting these issues, we can begin to develop a framework for ensuring that the resurrection of the dead is guided by principles of justice, compassion, and respect for the dignity of all life.

The Ethics of Restoring Identity

At the heart of any effort to resurrect the dead is the question of identity: What does it mean to bring someone back to life? Is it enough to recreate a person's physical body, or must we also restore their consciousness, memories, and personality? These questions highlight the complexity of defining what makes a person who they are.

1. Physical vs. Psychological Identity
   A person's identity encompasses both their physical form and their inner life—their thoughts, memories, emotions, and sense of self. If resurrection technologies can reconstruct a person's body but fail to restore their subjective experience, is the resurrected individual truly the same person? Ethical frameworks must grapple with this distinction, ensuring that efforts to restore life prioritize the holistic integrity of individuals.

2. The Continuity of Self
   Another challenge lies in the continuity of identity over time. If a person has been dead for centuries or millennia, can they still be meaningfully considered the same individual upon resurrection? Philosophers have long debated whether identity is tied to continuity of consciousness, physical form, or something more abstract, such as a soul. Resurrection efforts must confront these philosophical questions, striving to preserve the essence of the individual while acknowledging the potential for discontinuity.

3. Incomplete or Fragmented Data
   In many cases, the information needed to reconstruct a person may be incomplete or fragmentary. For example, if only partial genetic material or limited records of a person's life are available, any attempt to resurrect them will involve speculation and approximation. This raises ethical concerns about authenticity: Is it ethical to bring back a "version" of someone that may differ significantly from the original? Should resurrection only proceed when a high degree of accuracy can be guaranteed?

The Rights of the Resurrected

If scientific resurrection becomes possible, it will be essential to consider the rights and autonomy of those who are brought back to life. The resurrected will not be passive objects of scientific achievement but sentient beings with their own desires, needs, and perspectives. Ethical considerations must ensure that their dignity and agency are respected.

1. Consent and Autonomy

One of the most significant ethical dilemmas is the issue of consent. The deceased cannot provide informed consent for their resurrection, yet the act of restoring them to life imposes profound consequences on their existence. How do we determine whether someone would have wanted to be resurrected? Should their wishes during life take precedence, or should the potential benefits of resurrection override this concern?

Furthermore, once resurrected, individuals must be granted full autonomy over their lives. They should have the freedom to make their own decisions, including the choice to continue living or to return to death if desired.

2. Psychological and Emotional Challenges

Resurrection may impose significant psychological and emotional burdens on the resurrected. They may struggle to adapt to a world that has changed drastically since their time, facing cultural, technological, and existential disorientation. Ethical frameworks must include provisions for supporting the mental health and well-being of the resurrected, ensuring that their transition into the present world is as smooth and compassionate as possible.

3. Fair Treatment and Integration

Resurrected individuals must be treated as equals to the living, with full access to rights, resources, and opportunities. Society must address potential discrimination or stigmatization against the resurrected, ensuring that they are fully integrated into the fabric of the present without prejudice.

The Impact on Society and the Living

Resurrecting the dead will not only affect the individuals being restored but also have far-reaching consequences for society as a whole. These broader implications must be carefully considered to ensure that resurrection serves the greater good rather than creating new injustices or inequalities.

1. Resource Allocation
Bringing the dead back to life will require significant resources, including energy, time, and technology. Ethical questions arise about whether these resources might be better spent addressing the needs of the living, such as alleviating poverty, combating climate change, or curing diseases. How should society balance the pursuit of resurrection with its responsibilities to the living?

2. Overpopulation and Environmental Impact
If resurrection becomes widespread, it could lead to overpopulation and increased strain on the planet's resources. Ethical frameworks must consider the ecological consequences of resurrecting large numbers of people and develop strategies for sustainable population management.

3. Cultural and Historical Disruption
The resurrection of individuals from different time periods could disrupt cultural and historical narratives, challenging our understanding of identity and continuity as a species. How should society navigate the integration of individuals with vastly different worldviews, values, and knowledge? Should there be limits on who can be resurrected, or should all individuals be given an equal opportunity for restoration?

The Morality of Repairing the Past

The very act of resurrecting the dead is rooted in a desire to repair the past—to undo the injustices of death, loss, and historical erasure. While this goal is noble, it raises profound moral questions about humanity's role as stewards of the past.

1. Redemption and Justice

Resurrection offers the possibility of redeeming historical injustices, such as genocide, slavery, or systemic oppression, by restoring the lives and legacies of those who were wronged. However, this raises questions about how to prioritize whom to resurrect, as not all individuals can be restored simultaneously. Should priority be given to those who suffered the greatest injustices, or should resurrection be pursued without regard to historical context?

2. The Risk of Rewriting History

Resurrecting individuals or restoring historical events could unintentionally alter our understanding of history, creating new narratives that replace or overwrite existing ones. Ethical considerations must ensure that resurrection is conducted in a way that preserves the integrity of the past while acknowledging the complexity of historical truth.

3. The Burden of Playing God

The act of resurrecting the dead places humanity in a position of immense power, akin to the role traditionally attributed to divine beings. This raises questions about the morality of such an undertaking: Do humans have the right to reverse death, or is it an overreach that risks unintended consequences? Ethical frameworks must confront the tension between humanity's aspirations and its limitations, ensuring that resurrection is pursued with humility and caution.

Ethical Frameworks for Resurrection

Given the profound ethical challenges associated with resurrecting the dead, it is essential to develop robust frameworks to guide this endeavor. These frameworks should be grounded in principles of justice, compassion, equality, and respect for autonomy. Key considerations include:

1. Informed Decision-Making: Resurrection efforts should be guided by interdisciplinary collaboration, incorporating insights

from science, philosophy, ethics, and cultural studies to ensure a balanced approach.

2. Consent and Representation: Where possible, the wishes of the deceased should be respected, and efforts should be made to involve their descendants or representatives in the decision-making process.

3. Equity and Accessibility: Resurrection technologies must be made accessible to all, regardless of social, economic, or geographic status, to prevent the creation of new inequalities.

4. Sustainability and Responsibility: The pursuit of resurrection must be conducted in a way that minimizes harm to the living, the environment, and society as a whole.

Conclusion

The ethical implications of resurrecting the dead are as profound as the technological challenges it entails. While the possibility of restoring life offers immense hope and potential for healing, it also demands careful reflection on the nature of identity, the rights of the resurrected, and the impact on society and history. By addressing these ethical considerations with thoughtfulness and compassion, humanity can ensure that the pursuit of resurrection is not only a scientific achievement but also a moral triumph—one that respects the dignity of the past while building a brighter future for all.

Chapter 4: The Role of Advanced Technology

Section 1: Speculative Technology for Universal Resurrection

Subsection 1: Nanotechnology and Bioprinting for Physical Reconstitution

The physical restoration of individuals who have long since passed requires extraordinary advancements in science and engineering—technologies capable of reconstructing the body down to its smallest components. Enter nanotechnology and bioprinting, two groundbreaking fields that hold the promise of making physical reconstitution not only feasible but precise and efficient. These technologies offer pathways to rebuild the human body from the molecular level, restoring not just its outward form but its intricate biological functions.

In this subsection, we explore how nanotechnology and bioprinting could be employed to achieve the physical resurrection of the dead. By combining molecular-level manipulation with advanced 3D printing of biological structures, these technologies could enable the recreation of human bodies, organs, tissues, and even cellular-level details, paving the way for universal scientific resurrection.

Nanotechnology: Rebuilding Life from the Ground Up

Nanotechnology, the manipulation of matter at the atomic and molecular scale, lies at the heart of any effort to achieve physical reconstitution. At this scale, individual atoms and molecules can be rearranged to recreate structures with extraordinary precision. For the purposes of resurrection, nanotechnology offers the potential to rebuild the physical body from its most fundamental components.

1. Molecular Assemblers
   The concept of molecular assemblers—nanomachines capable of constructing complex structures atom by atom—has been a central idea in nanotechnology since it was first proposed by Eric Drexler

in the 1980s. In the context of resurrection, molecular assemblers could be programmed to rebuild a human body by precisely arranging atoms to recreate cells, tissues, and organs.

For example, given a genetic blueprint from DNA or even partial genetic material, molecular assemblers could reconstruct an entire organism by following the instructions encoded in the genome. This process would involve recreating not only the physical structure of the body but also its functional systems, such as the circulatory, nervous, and immune systems.

2. Repairing and Restoring Damaged Cells
Nanotechnology also offers the potential to repair or restore damaged biological material. For individuals whose remains have degraded over time, nanobots—microscopic robots designed to operate at the cellular level—could identify and repair damaged DNA, proteins, and other cellular components. This would enable the recovery of critical information needed for accurate physical reconstitution.

In cases where physical remains are too degraded to provide a complete genetic blueprint, nanotechnology could fill in the gaps by synthesizing missing components based on probabilistic models or data from related sources, such as descendants or environmental imprints.

3. Reconstructing the Brain and Neural Networks
One of the most challenging aspects of physical reconstitution is the restoration of the brain, the most complex organ in the human body. The brain's intricate network of neurons and synapses encodes memories, personality, and consciousness, making its accurate reconstruction essential for a true resurrection.

Nanotechnology could be used to map and rebuild neural networks at an atomic level, effectively recreating the individual's brain as it existed at the time of death. By combining this with advanced data retrieval methods, such as quantum archaeology, it may even be possible to restore aspects of the individual's memories and subjective experiences.

Bioprinting: Creating Life Layer by Layer

Bioprinting, a specialized form of 3D printing that uses biological materials to create living tissues and organs, is another critical technology for physical reconstitution. While nanotechnology operates at the molecular level, bioprinting works at the cellular and tissue levels, providing the framework and structure for rebuilding the human body.

1. Printing Organs and Tissues
Recent advancements in bioprinting have already demonstrated the ability to create functional human tissues, such as skin, cartilage, and blood vessels. Using bio-inks—specialized materials composed of living cells and biocompatible scaffolds—bioprinters can fabricate complex biological structures layer by layer.

For resurrection, bioprinting could be used to recreate entire organs, such as the heart, lungs, and kidneys, based on the individual's unique genetic blueprint. These organs could then be integrated into the reconstructed body, ensuring that all systems function harmoniously.

Beyond individual organs, bioprinting could also fabricate entire skeletal and muscular systems, providing the physical framework needed to support the body's movement and structural integrity.

2. Regenerating Skin and External Features
Bioprinting is particularly well-suited for recreating external features, such as skin, hair, and facial structures. By analyzing photographs, records, or even genetic data, bioprinters could restore an individual's outward appearance with remarkable accuracy. This ensures not only biological functionality but also the preservation of identity, allowing the resurrected individual to be recognized by loved ones and society.

3. Scalable Reproduction of Biological Systems
One of the key advantages of bioprinting is its scalability. Unlike traditional methods of organ transplantation or tissue regeneration, bioprinting can produce multiple copies of biological structures

simultaneously. This makes it possible to resurrect large numbers of individuals in parallel, significantly accelerating the process of universal scientific resurrection.

The Synergy of Nanotechnology and Bioprinting

While nanotechnology and bioprinting are powerful on their own, their true potential lies in their synergy. By combining the precision of nanotechnology with the structural capabilities of bioprinting, it becomes possible to achieve a level of accuracy and functionality that neither technology could accomplish alone.

1. Layer-by-Layer Assembly with Molecular Precision

Nanotechnology could be integrated into the bioprinting process to ensure molecular-level precision in the construction of tissues and organs. For example, nanobots could assist bioprinters by fine-tuning the placement of individual cells or molecules, ensuring that the resulting structures are identical to their original counterparts.

2. Dynamic Repair and Adaptation

Once the body has been reconstructed, nanotechnology could continue to play a role in maintaining and adapting the resurrected individual's physiology. For instance, nanobots could monitor and repair cellular damage in real time, ensuring that the body remains healthy and functional even in the face of environmental challenges.

3. Recreating Consciousness and Neural Integration

The synergy of nanotechnology and bioprinting is particularly important for the restoration of the brain and nervous system. While bioprinting provides the physical structure of the brain, nanotechnology could be used to map and integrate neural connections, ensuring that the resurrected individual's consciousness is fully restored.

Challenges and Ethical Considerations

The use of nanotechnology and bioprinting for physical reconstitution is not without its challenges. These include technical limitations, such as the difficulty of precisely mapping and recreating complex biological systems, as well as ethical concerns related to the authenticity and sustainability of resurrection.

1. Accuracy and Authenticity
How can we ensure that the reconstructed body and brain are faithful to the original individual? Small errors in the placement of cells or molecules could have significant consequences for identity and functionality. Efforts must be made to develop rigorous standards for accuracy, ensuring that resurrected individuals are as authentic as possible.

2. Resource Allocation
The process of physical reconstitution is likely to require significant resources, including energy, materials, and computational power. Ethical frameworks must address how these resources are allocated, ensuring that resurrection efforts do not come at the expense of pressing needs for the living.

3. Consent and Autonomy
As with any form of resurrection, the issue of consent remains a central ethical concern. How do we determine whether an individual would have wanted to be resurrected? Once resurrected, how do we ensure that they have full autonomy over their reconstructed body and life?

Conclusion

Nanotechnology and bioprinting represent two of the most promising technologies for achieving physical reconstitution as part of universal scientific resurrection. By working at both the molecular and cellular levels, these technologies offer the precision and scalability needed to restore the human body in all its complexity.

However, the path toward resurrection is not purely technical—it is also deeply ethical. As we develop the tools to reconstruct the past, we must ensure that our efforts are guided by principles of accuracy, compassion, and respect for the dignity of the individual. Together, nanotechnology and bioprinting bring us closer to a future where the boundaries of life and death are no longer insurmountable, offering humanity the possibility of repairing the past and reclaiming what was thought to be lost forever.

Subsection 2: Artificial Intelligence and Memory Reconstruction

One of the most profound challenges in the process of universal scientific resurrection is the restoration of memory. Memories are central to our identity—they define our relationships, our experiences, and our sense of self. Yet, when someone has passed away, their memories, seemingly ephemeral, are lost forever. To achieve true resurrection, it is not enough to reconstruct the physical body; we must also restore the intangible aspects of the individual, particularly their thoughts, emotions, and personal history.

Artificial intelligence (AI) emerges as a critical technology in this endeavor. Through its unparalleled ability to analyze, synthesize, and reconstruct information, AI holds the potential to access and recreate the memories of the deceased. By integrating advances in machine learning, neural networks, and data retrieval, AI systems could rebuild the lost cognitive and emotional landscapes that once defined a person's mind.

This subsection will explore how artificial intelligence might enable memory reconstruction, the tools and techniques involved, and the profound ethical implications of recreating someone's inner world.

The Role of Memory in Resurrection

To fully resurrect an individual, memory must be treated as an essential component of identity. A body without memories is merely a shell, while a mind without its history lacks coherence and continuity. Reconstructing memory is therefore a cornerstone of universal scientific resurrection.

1. Memory as a Continuum of Identity
Memories are the threads that weave a person's identity over time. They link the past to the present, forming the foundation of personality, knowledge, and emotional connections. Without memory, an individual loses their sense of who they are, where they come from, and how they relate to the world. Thus, memory reconstruction is essential to ensure that the resurrected individual is not merely a replica, but a continuation of the original person.

2. The Challenges of Memory Reconstruction
Memories are not stored in a single location or format; they are distributed across the brain in complex patterns of neural activity and synaptic connections. Over time, these patterns degrade as the brain ceases to function, making the task of recovering memories extraordinarily difficult. Additionally, memories are often incomplete, distorted, or influenced by emotions, raising questions about how to determine their authenticity during reconstruction.

AI Techniques for Memory Recovery and Reconstruction

Artificial intelligence offers a range of tools and techniques for addressing the challenges of memory reconstruction. By analyzing data from a variety of sources, simulating neural processes, and synthesizing fragmented information, AI systems could recreate the memories of the deceased with remarkable fidelity.

1. Data Retrieval from External Sources
While memories are intrinsically linked to the brain, traces of a person's life and experiences often exist outside their body. AI can

analyze this external data to reconstruct aspects of an individual's identity.

- Personal Archives: AI systems can process personal archives such as photographs, videos, journals, letters, and social media posts. By analyzing these materials, AI can infer key aspects of an individual's life, such as their relationships, preferences, and significant life events.

- Speech and Text Analysis: AI-powered natural language processing (NLP) can analyze recorded conversations, written texts, and digital communications to reconstruct a person's voice, thought patterns, and emotional tone. This data can be used to simulate how the individual might have expressed themselves.

- Environmental and Historical Records: AI can also draw upon broader environmental data, such as geographical records, historical events, or cultural artifacts, to contextualize an individual's experiences and enrich the reconstruction process.

2. Neural Network Simulations

Advances in AI-driven neural networks offer the possibility of simulating the brain's memory processes to recreate an individual's cognitive and emotional landscape.

- Brain Mapping and Reconstruction: Technologies such as functional MRI (fMRI) and connectomics are already being used to map the structure and activity of the brain. AI could analyze these maps to model the neural patterns associated with memory storage and retrieval, effectively simulating how the brain originally encoded memories.

- Machine Learning and Pattern Recognition: Using machine learning algorithms, AI could identify patterns within neural data or external sources that correspond to specific memories. These patterns could then be synthesized into a cohesive reconstruction of the individual's memory.

- Recreating Emotional Context: Memories are not just factual records but also emotional experiences. AI systems could integrate data on an individual's emotional reactions to recreate the feelings associated with specific memories, ensuring that the reconstructed mind is as rich and nuanced as the original.

3. Filling in the Gaps
In many cases, the available data may be incomplete or fragmented. AI excels at working with incomplete datasets, using predictive models and probabilistic reasoning to fill in the gaps.

- Probabilistic Memory Reconstruction: By analyzing patterns in the available data, AI could make informed predictions about missing details. For example, if a person's writings mention spending summers at a specific location, AI could infer details about their experiences based on historical and environmental data from that location.
- Extrapolation and Generalization: AI could also draw on broader patterns of human behavior, psychology, and cultural norms to extrapolate likely scenarios for missing memories. While these reconstructions may not be perfectly accurate, they could serve as reasonable approximations that preserve the essence of the individual's experiences.

Applications of AI in Resurrection

The ability to reconstruct memories using AI has profound implications for the process of resurrection. Below are some key applications of this technology:

1. Recreating Personal Identity
   AI-powered memory reconstruction ensures that the resurrected individual retains their unique identity, including their thoughts, feelings, and personal history. This makes the resurrection process more meaningful, as it restores not only the physical body but also the individual's sense of self.

2. Restoring Relationships
   Memories are the foundation of relationships. By reconstructing the memories of the deceased, AI can help restore their connections with loved ones, allowing families and communities to reunite with the person they once knew.

3. Preserving Cultural and Historical Legacies

Memory reconstruction can also be used to resurrect individuals who played significant roles in history, science, art, or culture. By restoring their memories and perspectives, AI can help preserve and enrich humanity's collective knowledge and heritage.

Ethical Considerations in Memory Reconstruction

The use of AI for memory reconstruction raises profound ethical questions that must be addressed to ensure that this technology is used responsibly and respectfully.

1. Authenticity and Accuracy

How can we ensure that reconstructed memories are authentic? If AI fills in gaps with predictive models or generalizations, there is a risk of creating memories that deviate from the original person's experiences. Ethical frameworks must establish standards for accuracy and transparency in the reconstruction process.

2. Consent and Privacy

Reconstructing someone's memories involves accessing deeply personal and private information. How do we obtain consent from individuals who are no longer alive? Should family members or descendants have the right to authorize memory reconstruction, and if so, under what conditions?

3. The Impact on the Resurrected

Restoring memories could have a profound psychological impact on the resurrected individual. They may struggle to reconcile their reconstructed memories with their new reality or experience distress if their memories are incomplete or inaccurate. Ethical guidelines must include provisions for supporting the mental and emotional well-being of the resurrected.

4. The Rights of the Living

Memory reconstruction may also affect the living, particularly if it involves the resurrection of individuals who have a complex or controversial legacy. How do we balance the rights of the living with the desire to restore the memories of the dead?

Conclusion

Artificial intelligence offers unparalleled potential for reconstructing the memories of the deceased, making it a cornerstone of universal scientific resurrection. Through advanced data analysis, neural network simulations, and predictive modeling, AI can recreate the cognitive and emotional landscapes that define an individual's identity.

However, the reconstruction of memory is not merely a technical challenge—it is also a profoundly ethical endeavor. As we develop the tools to restore the past, we must ensure that our efforts are guided by principles of authenticity, consent, and compassion. By addressing these challenges, AI can help humanity bridge the gap between life and death, creating a future where the memories of the past are never truly lost.

Section 2: Time Manipulation and Quantum Engineering

Subsection 1: Theoretical Approaches to Time Reversibility

The concept of time has long been a subject of fascination and debate among physicists, philosophers, and theologians. In our everyday experience, time appears to flow in a single direction—forward—marked by the irreversible progression of events and the inevitable decay of systems. Yet, at the deepest levels of nature, the laws of physics do not inherently prohibit the reversal of time. Theoretical approaches to time reversibility suggest that it might be possible to navigate this "arrow of time" not only to observe the past but to actively restore it. For the purposes of universal scientific resurrection, the ability to reverse time would provide a powerful mechanism for retrieving and repairing what has been lost.

This subsection explores the theoretical underpinnings of time reversibility, including concepts from thermodynamics, quantum mechanics, and spacetime physics. By examining these ideas, we can begin to envision how the future might repair the past—not metaphorically, but literally—through the reversal of time's flow.

Understanding the Arrow of Time

The "arrow of time" is a term used to describe the apparent one-way flow of time from the past to the future. This asymmetry is central to human experience, yet it is not explicitly mandated by the fundamental laws of physics. To understand how time might be reversed, it is essential to first consider the origins of this asymmetry.

1. Thermodynamic Arrow of Time

The most widely recognized explanation for the arrow of time arises from the <u>second law of thermodynamics</u>, which states that the entropy (or disorder) of a closed system tends to increase over time. This principle explains why certain processes—such as the breaking of a glass or the mixing of liquids—are irreversible in practice, even though the microscopic equations governing particle interactions are time-symmetric.

Time reversal at a macroscopic level would require the ability to decrease entropy, effectively "rewinding" the sequence of events that led to a particular state. While this seems infeasible in ordinary circumstances, advances in nanotechnology, quantum mechanics, and computational physics may one day enable targeted reductions in entropy, allowing specific systems to be restored to their earlier configurations.

2. Cosmological Arrow of Time

The expansion of the universe provides another basis for the arrow of time. As the universe continues to expand from its initial state in the Big Bang, the direction of time is marked by this outward growth. Some cosmological theories, however, propose scenarios in which the universe might eventually contract, reversing the cosmological arrow of time. If such a reversal occurs, it could open the door to large-scale time reversibility, offering a framework for restoring past events on a cosmic scale.

3. Psychological Arrow of Time

Our perception of time's flow is deeply tied to memory: we remember the past but not the future. This psychological arrow of time is a product of how the brain processes information, influenced by the thermodynamic and cosmological arrows. If time reversibility becomes technologically feasible, it may challenge our innate understanding of memory, causality, and the flow of events.

Quantum Mechanics and Time Reversibility

Quantum mechanics, the branch of physics that governs the behavior of particles on the smallest scales, offers some of the most intriguing insights into time reversibility. While classical physics often assumes a fixed and linear progression of time, quantum mechanics reveals a more nuanced and probabilistic picture of reality.

1. Time-Symmetric Equations

The fundamental equations of quantum mechanics, such as the Schrödinger equation, are time-symmetric, meaning they work equally well whether time flows forward or backward. This symmetry suggests that, at least on the quantum level, time reversal is theoretically possible.

In practice, however, the act of measurement and observation introduces an element of irreversibility, collapsing a particle's wavefunction into a definite state. Overcoming this obstacle would require new methods of manipulating quantum systems to "undo" measurements and restore earlier states.

2. Quantum Entanglement and Information Retrieval

Quantum entanglement, the phenomenon in which particles become interconnected in such a way that the state of one particle instantaneously affects the state of another, may hold the key to time reversibility. Some researchers have proposed that entangled particles could act as a kind of "memory" of past states, allowing information about earlier configurations of matter to be reconstructed.

For instance, quantum archaeology—a speculative field— envisions using entangled systems to trace the quantum imprints left behind by past events. By decoding these imprints, scientists could effectively "replay" the past, reconstructing individuals, objects, and environments with extraordinary precision.

3. Quantum Retrocausality

Some interpretations of quantum mechanics, such as the transactional interpretation, introduce the concept of retrocausality, in which effects can influence their causes. This idea challenges the conventional flow of time, suggesting that information might be sent backward across time intervals. If retrocausality can be harnessed, it could provide a mechanism for repairing past events or even resurrecting individuals by transmitting corrective information into the past.

Spacetime Physics and Temporal Manipulation

Einstein's theory of relativity revolutionized our understanding of time, treating it as a dimension intertwined with space to form a unified fabric known as spacetime. Relativity opens the door to theoretical approaches for manipulating time, offering potential frameworks for reversing or revisiting the past.

1. Closed Timelike Curves (CTCs)

One of the most fascinating implications of general relativity is the possibility of closed timelike curves (CTCs), which are paths through spacetime that loop back on themselves. CTCs are often associated with theoretical constructs such as rotating black holes (Kerr black holes) or wormholes, which could allow an individual or object to travel backward in time.

While CTCs remain speculative and are fraught with paradoxes—such as the "grandfather paradox," in which a time traveler might alter their own existence—they offer a tantalizing glimpse of how time might be manipulated to access and repair the past.

2. Reversing Local Time

Instead of reversing time on a universal scale, some theories focus on localized time reversal, in which specific regions of spacetime are "rewound" without disrupting the broader flow of time. This approach would involve creating highly controlled environments, perhaps using advanced energy fields or spacetime distortions, to isolate and reverse the temporal progression of specific systems.

For instance, a localized time reversal device might be used to restore a decayed biological body to its earlier, living state, providing a practical tool for resurrection.

3. Holographic Universe and Spacetime Archives
The holographic principle, a theoretical framework derived from string theory, suggests that all the information about a three-dimensional region of spacetime may be encoded on its two-dimensional boundary. If this principle holds true, the past states of the universe could be preserved as "holographic records" on the surface of spacetime. Advanced technologies could potentially access and decode these records, enabling the recreation of past events and individuals.

The Challenges and Paradoxes of Time Reversibility

While the theoretical approaches to time reversibility are compelling, they are not without significant challenges and paradoxes. These include:

1. The Problem of Causality
Reversing time introduces the risk of causal paradoxes, in which changes to the past alter the conditions that originally led to the present. Such paradoxes could disrupt the continuity of events, creating inconsistencies that are difficult to reconcile. Advanced theories must address how to navigate or resolve these paradoxes.

2. Energy and Entropy
Reversing time requires overcoming the natural tendency of systems to increase in entropy. This process would likely demand immense amounts of energy, raising questions about the feasibility and sustainability of large-scale time reversal.

3. Ethical Implications

The ability to reverse time raises profound ethical questions about the limits of human intervention in the past. Should we undo historical tragedies, even if it means altering the present? Should everyone be resurrected, or only specific individuals? These dilemmas must be carefully considered as theoretical concepts move closer to practical applications.

Conclusion

Theoretical approaches to time reversibility offer tantalizing possibilities for repairing the past and enabling universal scientific resurrection. From the time-symmetric equations of quantum mechanics to the spacetime distortions of general relativity, these ideas challenge our conventional understanding of time and open the door to revolutionary technologies.

While significant obstacles remain—both technical and philosophical—the exploration of time reversibility reminds us that the boundaries of possibility are far from fixed. As our understanding of time deepens, we may one day find ourselves capable of revisiting and restoring the past, fulfilling humanity's age-old dream of transcending the finality of death. Through the lens of time, the future may indeed be poised to repair the past.

Subsection 2: Bridging Quantum Physics with Resurrection Goals

Quantum physics, the study of the fundamental building blocks of reality, offers a revolutionary framework for understanding the universe at its most granular level. Unlike classical physics, which operates within the boundaries of determinism and predictability, quantum physics reveals a world of probabilities, entanglements, and hidden information. These principles, though abstract and often counterintuitive, have profound implications for the concept of universal scientific resurrection.

By bridging the principles of quantum physics with the ambitious goal of resurrecting the dead, we move closer to understanding how the information that constitutes a person—their physical body, consciousness, and memories—might be retrieved and restored. This subsection explores how quantum phenomena such as superposition, entanglement, and the nature of information can intersect with resurrection goals, offering novel pathways for bringing the past back to life.

The Quantum Nature of Information

In the realm of quantum physics, information is not merely an abstract concept—it is a fundamental aspect of reality. The study of quantum information suggests that the universe itself might be a vast repository of data, with every particle and interaction encoding details about its history. This insight is critical for resurrection goals, as it implies that the information needed to reconstruct an individual may not be lost, but rather dispersed or hidden at the quantum level.

1. The Principle of Information Conservation

One of the most important ideas in quantum physics is the conservation of information. According to this principle, information about the state of a physical system is never truly destroyed, even when the system undergoes dramatic changes. For example, the "no-hiding theorem" suggests that while information may become scrambled or inaccessible, it remains embedded in the universe.

This principle underpins the idea that the information needed to resurrect the dead—such as the arrangement of their atoms, the structure of their brains, and their unique memories—still exists in some form. The challenge lies in retrieving and decoding this information, a task that quantum physics may help us accomplish.

2. Quantum States as Carriers of Identity

In quantum mechanics, the state of a particle is described by a wavefunction, a mathematical representation that encodes all the probabilities of its possible behaviors. Extending this concept to larger systems, it is conceivable that a person's physical and mental state could be represented as an extremely complex wavefunction. If this wavefunction can be reconstructed, it may be possible to recreate the individual in their entirety.

The reconstruction of a person's quantum state would require precise knowledge of their physical configuration and mental processes at a specific point in time. Advances in quantum simulation and data retrieval could provide the tools needed to achieve this level of precision.

Quantum Entanglement and Resurrection

Quantum entanglement, a phenomenon in which the states of two or more particles become interconnected regardless of the distance between them, offers another intriguing avenue for resurrection goals. Entanglement suggests that information about a system can be distributed across multiple particles, creating a network of connections that persists even when the system is disrupted.

1. Entanglement as a "Memory" of the Past
When particles interact, they become entangled in such a way that their states are interdependent. This means that even if the particles are separated, their shared history is preserved in their entangled states. Some researchers have speculated that entanglement might serve as a kind of "memory" of past interactions, encoding information about the configuration of a system at earlier points in time.

If this principle can be extended to macroscopic systems, it could enable the retrieval of information about individuals and events from the past. By analyzing entangled particles and their correlations, scientists might reconstruct the physical and mental states of people who have long since passed away.

2. Quantum Archaeology
Building on the concept of entanglement, the speculative field of quantum archaeology envisions using quantum technologies to "dig up" the quantum imprints left behind by past events. These imprints, encoded in the entangled states of particles, could serve

as a blueprint for reconstructing individuals, including their bodies, memories, and consciousness.

While quantum archaeology remains highly theoretical, it represents a promising intersection of quantum physics and resurrection goals. By treating the universe as a vast archive of quantum information, this approach offers a pathway for restoring the past with unprecedented accuracy.

Superposition and the Reconstruction of Possibilities

Another key principle of quantum physics is **superposition**, the ability of particles to exist in multiple states simultaneously until they are observed or measured. Superposition challenges classical notions of reality, suggesting that the universe operates as a web of probabilities rather than fixed outcomes. This idea has profound implications for resurrection, particularly in scenarios where information about a person is incomplete or uncertain.

1. Reconstructing Probabilities
   In cases where the data required to reconstruct an individual is fragmented or degraded, superposition could provide a way to explore multiple possibilities simultaneously. For example, an AI system guided by quantum algorithms could model various probable configurations of a person's physical and mental state, based on the available evidence.

   By iterating through these possibilities, the system could identify the most likely reconstruction of the individual, ensuring that the resurrected person is as accurate as possible. This approach may not fully eliminate uncertainty, but it offers a scientifically grounded method for approximating lost information.

2. Parallel Realities and Quantum Multiverses
   Some interpretations of quantum mechanics, such as the Many-Worlds Interpretation, propose the existence of parallel realities or multiverses, in which every possible outcome of a quantum event is realized in a separate branch of the universe. If these parallel

realities are accessible, they may contain alternate versions of individuals who have died in our universe.

Resurrection efforts could leverage this concept by "borrowing" information from parallel realities to fill in gaps or even by transferring individuals from one branch of the multiverse to another. While this idea remains speculative, it highlights the extraordinary potential of quantum physics to expand the boundaries of what is possible.

Quantum Computing and Resurrection

Quantum computing, an emerging technology that harnesses the principles of quantum mechanics, could play a pivotal role in achieving resurrection goals. Unlike classical computers, which process information using binary bits (0s and 1s), quantum computers use qubits, which can exist in superposition. This allows quantum computers to perform complex calculations at speeds far beyond the capabilities of classical systems.

1. Simulating the Past

Quantum computers could be used to simulate the past with extraordinary detail, reconstructing the physical and mental states of individuals based on quantum data. By processing vast amounts of information simultaneously, a quantum computer could model the interactions and configurations that defined a person's life, effectively "replaying" their existence in a virtual environment.

These simulations could serve as a preliminary step toward physical resurrection, providing a blueprint for recreating the individual in the real world.

2. Decoding Quantum Information

Retrieving and decoding quantum information about the past requires immense computational power, as the data is often highly entangled and dispersed across the universe. Quantum computers, with their ability to process entangled states and perform complex optimizations, are uniquely suited to this task.

By analyzing quantum correlations and reconstructing wavefunctions, quantum computers could unlock the information needed to restore the dead, bridging the gap between theoretical physics and practical resurrection.

Challenges and Philosophical Considerations

While the principles of quantum physics offer exciting possibilities for resurrection, they also raise significant challenges and philosophical questions:

1. Accuracy and Authenticity

How can we ensure that the reconstructed individual is truly identical to the original? Quantum processes are inherently probabilistic, meaning that some degree of uncertainty may persist. Ethical guidelines must address the implications of resurrecting individuals who may differ in subtle ways from their original selves.

2. The Nature of Consciousness

Quantum physics challenges our understanding of consciousness, raising questions about whether it can be fully reconstructed. Is consciousness a purely physical phenomenon, or does it involve non-physical elements that cannot be captured by quantum data?

3. Ethical Boundaries

The ability to manipulate quantum information raises ethical concerns about privacy, consent, and the limits of human intervention in nature. Should we resurrect everyone, or only those who meet certain criteria? How do we balance the rights of the living with the desire to restore the dead?

Conclusion

Quantum physics provides a fertile ground for exploring the possibility of universal scientific resurrection. From the conservation of information to the mysteries of entanglement and superposition, the principles of quantum mechanics offer powerful tools for retrieving and restoring the past.

However, the journey from theory to practice is fraught with challenges, both technical and philosophical. As we continue to deepen our understanding of quantum physics, we must approach the goal of resurrection with humility, rigor, and ethical responsibility. By bridging the quantum world with resurrection goals, we take a bold step toward a future where the boundaries between life and death are no longer insurmountable.

Part III: Toward a Universal R&D Program

Chapter 5: Designing an R&D Program for Resurrection

Section 1: Key Goals and Milestones

Subsection 1: Mapping Hidden Niches and Collecting Data

The foundation of universal scientific resurrection lies in the collection of data—the retrieval of every fragment of information necessary to reconstruct individuals, their lives, and the environments they inhabited. Humanity's past, though seemingly lost to the passage of time, leaves behind traces: physical remnants, genetic material, artifacts, and even subtle imprints on the environment. The challenge lies in mapping hidden niches where this information resides and developing advanced methods to extract, preserve, and interpret it.

This subsection explores the process of uncovering hidden data sources, from biological remains to environmental archives, and examines how emerging technologies may enable us to recover the rich tapestry of information required for scientific resurrection. By leveraging cutting-edge tools and interdisciplinary approaches, we can begin to assemble the puzzle of the past, piece by piece.

The Importance of Hidden Niches in Resurrection Efforts

Hidden niches represent the diverse locations and mediums where information about the past is stored, often in forms that are difficult to access or interpret. These niches hold the key to reconstructing the physical, mental, and cultural dimensions of individuals who have passed away.

1. Biological Niches
   Biological material, such as DNA, is one of the most critical sources of information for resurrection. DNA contains the genetic blueprint of an individual, encoding details about their physical form, predispositions, and even aspects of their personality. However, as time passes, biological material degrades, and much of it becomes embedded in hard-to-reach environments.

- <u>Fossilized Remains</u>: Bones, teeth, and other fossilized remains can preserve genetic material for thousands or even millions of years. Advances in paleogenomics—extracting and sequencing DNA from ancient remains—have already allowed researchers to reconstruct the genomes of extinct species such as Neanderthals. Similar techniques could be applied to human resurrection, provided that sufficient genetic material can be recovered.

- <u>Soft Tissue and Preserved Bodies</u>: Exceptional cases, such as mummified bodies or individuals preserved in ice or peat bogs, offer more intact biological samples. These specimens provide not only DNA but also insights into the individual's diet, health, and lifestyle.

- <u>Environmental Traces</u>: Even when remains are not directly accessible, traces of biological material (such as hair, skin cells, or bodily fluids) may be preserved in the environment, offering alternative pathways for data collection.

2. <u>Environmental Niches</u>

Beyond biological material, the environment itself serves as a vast archive of information about the past. By examining the physical and chemical signatures left behind by individuals and their activities, we can reconstruct key aspects of their lives.

- <u>Sedimentary Records</u>: Layers of sediment in lakes, oceans, and other geological formations can preserve traces of human activity, such as pollen, charcoal, or isotopic signatures. These records provide clues about the environment in which an individual lived, their interactions with nature, and the broader context of their existence.

- <u>Ice Cores</u>: Ice cores drilled from glaciers and polar regions contain bubbles of ancient air, preserving a record of atmospheric conditions, volcanic activity, and even human pollution. By analyzing these cores, researchers can infer details about the climate and environmental conditions that shaped the lives of past generations.

- <u>Microbial Niches</u>: The human body is home to a unique microbiome—an ecosystem of bacteria, viruses, and other microorganisms. Even after death, remnants of this microbiome may persist in soil or other environments, offering a potential source of information about an individual's health, diet, and lifestyle.

3. Cultural and Historical Niches

Cultural artifacts, written records, and oral traditions provide another layer of data for resurrection efforts. While these sources may not directly encode the physical attributes of an individual, they offer valuable insights into their beliefs, values, and relationships.

- Textual Records: Diaries, letters, and other written records capture the thoughts, emotions, and experiences of individuals in their own words. Advances in artificial intelligence (AI) could enable the analysis and synthesis of these records to reconstruct aspects of a person's cognitive and emotional life.
- Artifacts: Tools, clothing, and other personal belongings provide clues about an individual's daily life, social status, and cultural context. High-resolution imaging and material analysis techniques could be used to extract detailed information from these items.
- Oral Histories: Traditions passed down through generations may preserve stories and memories of individuals who have long since passed away. By digitizing and analyzing these narratives, researchers can uncover additional dimensions of the past.

Technologies for Mapping and Collecting Data

Mapping hidden niches and collecting data for resurrection requires a multidisciplinary approach, combining techniques from biology, archaeology, geology, and computer science. Several emerging technologies are particularly well-suited to this task.

1. Remote Sensing and Imaging

Advanced imaging technologies allow researchers to explore hidden niches without disturbing them, providing non-invasive methods for data collection.

- LIDAR (Light Detection and Ranging): LIDAR uses laser pulses to create detailed 3D maps of landscapes, even through dense vegetation. This technology has already been used to uncover ancient cities and burial sites, and it could play a key role in identifying locations where human remains or artifacts are buried.

- Ground-Penetrating Radar (GPR): GPR uses electromagnetic waves to detect subsurface features, such as graves or buried structures. By mapping these features, researchers can locate potential sources of resurrection data.

- Spectroscopy: Techniques such as Raman spectroscopy and infrared spectroscopy can analyze the chemical composition of materials, identifying traces of biological or cultural significance.

2. Genetic Sequencing and Reconstruction

The recovery and analysis of genetic material are central to resurrection efforts. Advances in genetic sequencing technologies have dramatically increased the speed and accuracy of DNA analysis.

- Next-Generation Sequencing (NGS): NGS allows researchers to sequence entire genomes from small, degraded samples, making it possible to recover genetic information from ancient remains.

- Genome Editing: Tools such as CRISPR-Cas9 enable researchers to repair damaged DNA or synthesize missing segments, providing a pathway for reconstructing complete genomes even when the original material is incomplete.

- Epigenetic Analysis: Beyond DNA, epigenetic markers—chemical modifications to the genome—provide insights into an individual's health, environment, and experiences. By analyzing these markers, researchers can create a more comprehensive picture of the individual.

3. Artificial Intelligence and Data Integration

AI plays a critical role in mapping and integrating data from diverse sources, enabling researchers to identify patterns and reconstruct missing information.

- Data Mining: AI algorithms can sift through vast amounts of historical, biological, and environmental data to identify relevant information about individuals and their contexts.

- <u>Pattern Recognition</u>: Machine learning models can identify correlations between different data types, such as genetic sequences and cultural artifacts, helping researchers piece together the puzzle of the past.

- <u>Simulation</u>: AI-powered simulations can recreate historical environments, allowing researchers to test hypotheses and refine their understanding of how individuals lived and interacted with their surroundings.

Challenges and Ethical Considerations

Mapping hidden niches and collecting data for resurrection is a monumental task, fraught with technical and ethical challenges.

1. Data Degradation
Over time, biological and environmental materials degrade, making it difficult to recover complete information. Researchers must develop innovative techniques to extract and interpret data from highly fragmented or contaminated samples.

2. Authenticity and Accuracy
How do we ensure that the data we collect accurately represents the individual or event in question? The integration of multiple data sources increases the risk of introducing errors or biases.

3. Privacy and Consent
The act of resurrecting individuals raises questions about privacy and consent. Do the dead have a right to determine how their remains and memories are used? What responsibilities do researchers have toward the individuals they seek to restore?

4. Resource Allocation
Mapping hidden niches and collecting data requires significant resources, including time, energy, and funding. Ethical frameworks must address how these resources are allocated, ensuring that resurrection efforts do not detract from addressing pressing issues in the present.

Conclusion

Mapping hidden niches and collecting data is the first step toward realizing the vision of universal scientific resurrection. By uncovering the traces of the past that lie buried in biological, environmental, and cultural archives, we can begin to reconstruct the lives of those who have come before us.

This process will require not only advanced technologies but also a deep commitment to ethical principles, ensuring that our efforts to repair the past are guided by respect for the dignity of the individuals we seek to resurrect. As we embark on this journey, we move closer to a future where no life is truly lost, and the boundaries of time and memory are transcended.

Subsection 2: Developing Tools for Quantum and Biological Reconstruction

The ambitious goal of <u>universal scientific resurrection</u> depends on the development of advanced tools capable of reconstructing both the physical and informational aspects of individuals who have long since passed. This monumental task requires bridging two complementary domains: <u>quantum reconstruction</u>, which focuses on retrieving and organizing the encoded information about a person, and <u>biological reconstruction</u>, which involves using that information to recreate the physical body and its intricate systems.

The tools for these processes must operate at the cutting edge of technology, incorporating breakthroughs in quantum mechanics, nanotechnology, synthetic biology, and artificial intelligence. This subsection explores the key technologies and methodologies that could make quantum and biological reconstruction feasible and discusses the challenges and ethical considerations associated with their development.

<u>Quantum Reconstruction: Retrieving the Information of the Past</u>

At the heart of quantum reconstruction lies the understanding that the universe operates as a vast repository of information.

Theoretical and experimental physics suggest that the information about every physical system, including human beings, is never truly lost but distributed and encoded in the environment. Developing tools to retrieve this quantum information is essential for the resurrection process.

1. Quantum Memory and Information Retrieval
Quantum memory refers to the storage of quantum states, which encode information at the smallest scales. Tools for quantum memory retrieval could allow us to extract lost or hidden information about past physical systems.

- Quantum Imprinting: As particles interact, they leave behind subtle quantum imprints in the environment. These imprints could serve as a record of the configurations and states of objects and individuals. Developing tools like quantum sensors or quantum archaeometry could allow scientists to detect and reconstruct these imprints, much like piecing together a shattered glass.
- Quantum Decoders: Quantum computers, capable of processing entangled and superposed states, could be used to decode the complex patterns of quantum information stored in entangled particles. These decoders would serve as the foundation for retrieving the complete quantum state of an individual or system.

2. Quantum Simulation of the Past
Once quantum information is retrieved, it must be organized into a coherent model of the individual. This requires the use of quantum simulation tools that can recreate the physical and informational state of a person from their quantum "blueprint."

- Wavefunction Reconstruction: An individual's quantum state, including their physical structure and mental processes, could be described by a highly complex wavefunction. Quantum computers can simulate these wavefunctions, effectively creating a virtual model of the individual.
- Entanglement Mapping: Tools for mapping quantum entanglement could provide insights into how particles were connected in the past, revealing the interdependent relationships

between an individual's physical components and their environment.

Biological Reconstruction: Rebuilding the Physical Body

Once the informational foundation is established through quantum reconstruction, the next step is to use this data to recreate the physical body. Biological reconstruction involves the development of tools and techniques to rebuild cells, tissues, organs, and even entire organisms.

1. Nanotechnology for Molecular Reconstruction
Nanotechnology operates at the atomic and molecular scale, making it an essential tool for rebuilding biological structures with precision.

- Molecular Assemblers: These nanomachines are designed to manipulate individual atoms and molecules, assembling them into complex structures. Molecular assemblers could be used to reconstruct cells, tissues, and even entire organs based on the quantum blueprint retrieved during the quantum reconstruction phase.
- DNA Repair and Synthesis: For individuals whose DNA is incomplete or damaged, nanotechnology could repair existing genetic material or synthesize missing segments. This would ensure the accuracy of the reconstructed genome, which serves as the foundation for biological regeneration.

2. Bioprinting and Regeneration
Bioprinting, a specialized form of 3D printing that uses biological materials, offers a scalable and efficient method for rebuilding the body.

- Organ Printing: Advances in bioprinting have already enabled the creation of functional tissues and small-scale organs. Future tools could scale up this technology, allowing scientists to print entire organ systems customized to the individual's genetic blueprint.

- Whole-Body Bioprinting: The ultimate goal of bioprinting is the ability to recreate an entire human body. This would involve printing not just organs but also bones, muscles, skin, and other tissues, all integrated into a cohesive and functional system.
- Stem Cell Engineering: Stem cells, which have the ability to differentiate into any cell type, could be programmed using the retrieved genetic data to regenerate specific tissues. Tools for stem cell engineering would play a critical role in rebuilding the body from the ground up.

3. Neural Reconstruction and Consciousness Integration
Rebuilding the physical body is only part of the resurrection process. The brain, with its intricate network of neurons and synapses, must also be reconstructed to restore the individual's memories, personality, and consciousness.

- Neural Bioprinting: Tools for bioprinting neural tissues could allow scientists to recreate the structure of the brain, including its vast network of connections. This would serve as the physical foundation for restoring cognitive function.
- Memory Encoding: Once the brain is physically reconstructed, the memories and personality of the individual must be re-encoded. This could be achieved using neural interface technologies that integrate quantum information into the brain's reconstructed neural network.
- Consciousness Simulators: Before physical integration, virtual simulations of an individual's consciousness could be tested and refined, ensuring that the reconstructed brain accurately reflects their original personality and experiences.

Synergizing Quantum and Biological Tools

The true potential of resurrection lies in the synergy between quantum and biological reconstruction tools. By integrating these technologies, we can create a seamless process that bridges the informational and physical aspects of the individual.

1. Quantum-Informed Bioprinting

Quantum reconstruction provides the blueprint for biological reconstruction. Tools for quantum-informed bioprinting would translate the retrieved quantum data into physical structures, ensuring that the reconstructed body is an exact replica of the original.

2. Dynamic Feedback Systems

The reconstruction process could involve dynamic feedback systems, in which quantum simulations are continuously updated based on biological progress. For example, as organs are printed and tested, the quantum model could refine its predictions, ensuring that the final product is both accurate and functional.

3. Integrated AI Systems

Artificial intelligence would serve as the bridge between quantum and biological tools, coordinating the retrieval, interpretation, and application of data. AI algorithms could optimize the reconstruction process, reducing errors and ensuring consistency across all levels of the individual's restoration.

Challenges in Developing Reconstruction Tools

The development of tools for quantum and biological reconstruction is a daunting endeavor, fraught with technical, ethical, and philosophical challenges.

1. Technical Limitations

- Precision: Both quantum and biological reconstruction require an unprecedented level of precision, as even small errors could have significant consequences for the final outcome.
- Scalability: Tools must be capable of operating at both microscopic and macroscopic scales, from manipulating individual atoms to reconstructing entire organs.

2. Ethical Considerations

- Authenticity: How do we ensure that the reconstructed individual is truly identical to the original?

- Consent: What ethical frameworks should govern the resurrection of individuals who cannot provide consent?

3. Resource Allocation
   The development and deployment of reconstruction tools will require significant resources, including energy, materials, and computational power. Society must grapple with how to prioritize these efforts alongside other global challenges.

Conclusion

The development of tools for quantum and biological reconstruction represents one of the most ambitious undertakings in human history. By retrieving the informational essence of individuals through quantum technologies and rebuilding their physical forms using advanced biological techniques, we can move closer to the vision of universal scientific resurrection.

However, the journey toward this goal is as much a philosophical and ethical challenge as it is a technical one. As we develop these tools, we must ensure that they are guided by principles of compassion, respect, and humility. In doing so, we can work toward a future where the boundaries between life and death are no longer insurmountable, and the past can be repaired with the tools of the future.

Section 2: Collaborating Across Disciplines

Subsection 1: Integrating Physics, Biology, and Philosophy

The quest for universal scientific resurrection is not merely a technical challenge—it is an interdisciplinary endeavor that merges the realms of physics, biology, and philosophy. Each of these fields offers unique insights into the nature of existence, identity, and the boundaries of life and death. Physics provides the foundational understanding of the universe's laws and the conservation of information. Biology offers the tools and frameworks for reconstructing life. Philosophy addresses the profound ethical, metaphysical, and existential questions that arise along the way.

To achieve the goal of resurrecting the dead, these three domains must work in harmony, creating a unified framework that transcends their individual boundaries. This subsection explores how physics, biology, and philosophy intersect in the pursuit of resurrection, and how their integration is essential for addressing the technical and ethical challenges of restoring the past.

<u>Physics: The Foundation of Reality and Information Conservation</u>

Physics forms the bedrock of our understanding of the universe. The principles of quantum mechanics, relativity, and thermodynamics provide the theoretical tools for addressing the fundamental question: <u>Is resurrection scientifically possible?</u>

1. <u>The Conservation of Information</u>
   One of the central principles of physics is the <u>conservation of information</u>, which suggests that the information about any physical system is never destroyed but remains encoded in the universe. This principle is crucial for resurrection efforts, as it implies that the data necessary to reconstruct an individual—down to the arrangement of their atoms and the configuration of their memories—still exists, albeit in a highly dispersed form.

   - <u>Black Hole Information Paradox</u>: The debate surrounding whether black holes destroy information has led to breakthroughs in understanding how information might be preserved in the universe. Insights from this research can inform how we retrieve data about individuals from the distant past.
   - <u>Holographic Principle</u>: The holographic principle, which posits that all the information contained within a volume of space can be encoded on its boundary, suggests that the universe itself may store a "record" of past events. Advanced tools could one day access this encoded information for resurrection purposes.

2. Quantum Mechanics and Time Reversibility

Quantum physics challenges classical notions of time and causality, opening the door to the possibility of reversing or reconstructing past events.

- Wavefunction Reconstruction: The quantum state of a system encodes its entire history, allowing for the possibility of recreating past configurations.
- Entanglement and Memory: Quantum entanglement connects particles across space and time, potentially preserving information about past interactions. Tools for mapping entanglement could reveal the hidden "memories" of individuals and their environments.

3. Spacetime and Relativity

Einstein's theory of general relativity treats time as a dimension intertwined with space, suggesting that past, present, and future coexist within the fabric of spacetime.

- Closed Timelike Curves (CTCs): Theoretical constructs such as CTCs could allow for the retrieval of information from the past by navigating spacetime in novel ways.
- Cosmological Preservation: The expanding universe and the flow of entropy may still preserve the imprints of past states in ways that advanced technologies could exploit.

Biology: Reconstructing the Building Blocks of Life

While physics provides the theoretical framework, biology offers the practical tools and processes for recreating life. Resurrection requires a deep understanding of how to reconstruct not only the human body but also the mind, which includes memory, personality, and consciousness.

1. Genetic Reconstruction

DNA is the blueprint for life, containing the instructions for building and maintaining a human being. Advances in genetics and synthetic biology are paving the way for reconstructing individuals from even the smallest fragments of biological material.

- Genome Synthesis: Tools such as CRISPR-Cas9 and other genome-editing technologies enable the repair or synthesis of genetic material, allowing researchers to recreate full genomes even from degraded samples.
- Epigenetics: Beyond DNA, epigenetic markers provide additional layers of information about an individual's traits, health, and experiences. Reconstructing these markers is essential for creating a faithful replica of the original person.

2. Cellular and Tissue Regeneration
Rebuilding the body requires the regeneration of cells, tissues, and organs, all integrated into a functioning system.

- Stem Cell Technology: Stem cells offer the ability to regenerate any type of tissue, providing a foundation for rebuilding the body from the ground up.
- Bioprinting: Advances in 3D bioprinting allow for the creation of tissues and organs, which can be assembled into a complete biological system.

3. Neurobiology and Consciousness
Reconstructing the brain is one of the most challenging aspects of resurrection, as it involves not only physical structures but also the intangible aspects of identity.

- Brain Mapping: Technologies such as connectomics aim to map the brain's neural connections, providing a "wiring diagram" that can be used to recreate its structure.
- Memory Encoding: Memories are stored as patterns of neural activity and synaptic connections. Tools for decoding and re-encoding these patterns are essential for restoring an individual's sense of self.
- Artificial Consciousness Integration: Before physical reconstruction, virtual simulations of consciousness could serve as a testing ground, ensuring that the recreated individual is accurate and coherent.

Philosophy: Addressing the Ethical and Metaphysical Questions

The scientific pursuit of resurrection raises profound philosophical questions about the nature of identity, the meaning of life, and the limits of human intervention. Philosophy provides a framework for addressing these questions and ensuring that resurrection efforts are guided by ethical principles.

1. The Nature of Identity

What makes a person who they are? Is identity defined by their physical body, their memories, or something more intangible, such as their soul or consciousness?

- Continuity vs. Replication: Is a resurrected individual truly the same person, or merely a replica? Philosophical debates about the continuity of identity must inform the criteria for successful resurrection.
- The Mind-Body Problem: The relationship between the physical brain and consciousness remains unresolved. Understanding this relationship is critical for determining whether consciousness can be faithfully restored.

2. The Ethics of Resurrection

The act of bringing the dead back to life raises significant ethical questions that must be addressed before resurrection technologies are deployed.

- Consent: Can someone who has died give consent to be resurrected? Should their family or descendants have the right to make this decision?
- Justice and Equality: Who should be resurrected, and who decides? Ensuring that resurrection efforts are equitable and inclusive is a key ethical challenge.
- Impact on the Living: How will the resurrection of the dead affect the living, particularly in terms of resources, relationships, and societal dynamics?

3. The Meaning of Life and Death
Resurrection challenges traditional beliefs about the finality of death and the meaning of life.

- Religious Perspectives: Many religious traditions have their own views on resurrection, the afterlife, and the sanctity of death. Scientific resurrection must navigate these perspectives with sensitivity and respect.
- Existential Implications: If death is no longer permanent, how does this change our understanding of what it means to live a meaningful life?

The Need for Integration

The integration of physics, biology, and philosophy is not optional—it is essential. Each field addresses a different facet of the resurrection process, and their combined insights are necessary for overcoming both technical and ethical challenges.

1. Interdisciplinary Collaboration
The development of resurrection technologies requires scientists, philosophers, ethicists, and theologians to work together, ensuring that technical advances are aligned with ethical principles.

2. A Unified Framework
By combining the theoretical insights of physics, the practical tools of biology, and the moral guidance of philosophy, we can create a coherent framework for resurrection that is both scientifically rigorous and ethically sound.

Conclusion

Integrating physics, biology, and philosophy is the cornerstone of universal scientific resurrection. Physics provides the theoretical foundation, biology offers the practical tools, and philosophy ensures that the process is guided by ethical and existential considerations. Together, these disciplines form a holistic approach to repairing the past and transcending the boundaries of life and death.

As we move closer to realizing the dream of resurrection, we must remain mindful of the profound questions it raises and the responsibilities it entails. Only by integrating knowledge, technology, and ethics can we ensure that the future's efforts to repair the past are both meaningful and just.

Subsection 2: Creating a Global Research Network for Resurrection

The pursuit of <u>universal scientific resurrection</u> is a grand vision that transcends the boundaries of individual disciplines, institutions, and nations. It is an endeavor that requires the collaboration of the brightest minds across the globe, the pooling of vast resources, and the integration of diverse cultural, scientific, and philosophical perspectives. The creation of a <u>global research network for resurrection</u> is essential to coordinate and advance the interdisciplinary efforts needed to bring this ambitious goal to fruition.

This subsection explores the structure, objectives, and challenges of establishing such a network. It also highlights the importance of fostering international collaboration, ethical oversight, and public engagement to ensure that resurrection research is conducted responsibly and inclusively.

The Vision for a Global Research Network

A global research network for resurrection would serve as a centralized yet collaborative system for coordinating efforts across various domains, including physics, biology, computer science, and ethics. Its overarching mission would be to unify the scientific and philosophical efforts required to develop the technologies and frameworks necessary for resurrecting the dead.

1. A Multidisciplinary Approach
   Resurrection is an inherently interdisciplinary challenge, requiring expertise from a wide range of fields:

- Physics: To explore time reversibility, quantum information retrieval, and the principles of spacetime reconstruction.
- Biology: To focus on genetic reconstruction, cellular regeneration, and the recreation of complex biological systems.
- Artificial Intelligence and Data Science: To process and integrate vast amounts of historical, genetic, and quantum data.
- Philosophy and Ethics: To address questions of identity, consent, and the societal implications of resurrection.
- Cultural Studies and History: To ensure that resurrection efforts honor the cultural and historical contexts of individuals and communities.

A global network would foster collaboration between these disciplines, enabling researchers to share insights, tools, and methodologies.

2. Research Hubs and Specializations

The network could be structured around a series of interconnected research hubs, each specializing in a particular aspect of resurrection science. For example:

- Quantum Information Retrieval Centers: Focused on developing tools to recover and decode information about the past.
- Biological Reconstruction Laboratories: Dedicated to advancing techniques for genome editing, tissue regeneration, and neural reconstruction.
- Ethics and Philosophy Committees: Responsible for developing ethical guidelines and addressing philosophical questions.
- Historical and Cultural Archives: Tasked with preserving and analyzing historical records and artifacts to inform the reconstruction of past lives.

These hubs would operate independently but collaborate through a shared digital infrastructure, ensuring that progress in one area informs and accelerates advancements in others.

3. An Open and Inclusive Framework

The global research network must be inclusive, welcoming contributions from scientists, philosophers, and ethicists from diverse cultural and geographical backgrounds. This inclusivity is essential to ensure that resurrection efforts reflect the values and perspectives of humanity as a whole, rather than being dominated by a single nation or ideology.

Objectives of the Global Research Network

The network would pursue a series of interconnected objectives, each contributing to the overarching goal of universal scientific resurrection.

1. Mapping and Preserving Global Data

The first step in resurrection is the collection and preservation of data about the past. The network would coordinate efforts to:

- Identify and map hidden niches where biological, quantum, and cultural information is stored.
- Develop technologies for retrieving and preserving this information, such as advanced imaging tools, quantum processors, and biological data banks.
- Create a global archive of historical, genetic, and environmental data, ensuring that it is accessible to researchers worldwide.

2. Developing Resurrection Technologies

The network would fund and support research into the tools and techniques required for resurrection, including:

- Quantum reconstruction tools, such as quantum computers and entanglement mapping devices.
- Biological reconstruction technologies, such as bioprinters, stem cell engineering platforms, and neural regeneration systems.
- AI-driven simulation and modeling systems for reconstructing individuals and their environments.

3. Ethical and Philosophical Research

Resurrection raises profound ethical and philosophical questions that must be addressed proactively. The network's committees would work to:

- Develop ethical guidelines for resurrection, including principles of consent, equity, and justice.
- Explore the philosophical implications of resurrection, such as the nature of identity and the meaning of life and death.
- Engage with religious, cultural, and philosophical traditions to ensure that resurrection efforts are respectful and inclusive.

4. Public Engagement and Education

The success of resurrection research depends on public support and understanding. The network would prioritize:

- Educating the public about the scientific and ethical aspects of resurrection.
- Engaging with communities to address concerns and incorporate their perspectives.
- Building trust through transparency and accountability.

Challenges in Building a Global Research Network

Creating a global research network for resurrection is a daunting task, and several challenges must be addressed to ensure its success.

1. Coordination Across Borders

International collaboration is often hindered by political, economic, and cultural differences. The network must navigate these challenges by fostering trust and mutual respect among participating nations and institutions.

2. Funding and Resource Allocation

Resurrection research is resource-intensive, requiring significant investments in technology, infrastructure, and personnel. The network must develop innovative funding models, such as public-private partnerships, to secure the necessary resources.

3. Ethical and Philosophical Disputes
The ethical and philosophical implications of resurrection are complex and often controversial. The network must create a framework for addressing disagreements and ensuring that research is guided by shared values.

4. Equity and Accessibility
Resurrection efforts must be accessible to all, rather than being limited to the wealthy or privileged. The network must prioritize equity and inclusivity in its research and implementation.

Case Study: A Hypothetical Global Initiative

To illustrate how a global research network for resurrection might function, consider the following hypothetical example:

The Resurrection Research Alliance (RRA)
The RRA is an international organization with member institutions from over 50 countries. It operates through a decentralized network of research hubs, each specializing in a different aspect of resurrection science.

- Headquarters: The RRA's central headquarters coordinates the network's activities and ensures that research aligns with shared goals and ethical principles.
- Regional Hubs: Each region has its own hub, such as the Quantum Information Center in Switzerland, the Biological Reconstruction Institute in Japan, and the Ethics and Philosophy Committee in South Africa.
- Global Archive: The RRA maintains a digital archive of historical, genetic, and environmental data, accessible to researchers worldwide.

The RRA's projects include developing quantum retrieval tools, designing AI-driven bioprinters, and hosting public forums to discuss the ethical implications of resurrection.

The Role of Technology in the Network

Modern technology will play a crucial role in enabling the global research network to function effectively.

1. Digital Collaboration Platforms
   Advanced digital platforms would allow researchers from around the world to collaborate in real time, sharing data, insights, and methodologies.

2. AI-Powered Coordination
   Artificial intelligence could optimize the network's operations by analyzing research progress, identifying synergies between projects, and allocating resources efficiently.

3. Blockchain for Transparency
   Blockchain technology could be used to ensure transparency and accountability, tracking funding, research outcomes, and ethical compliance.

Conclusion

The creation of a global research network for resurrection represents a critical step toward achieving the dream of universal scientific resurrection. By fostering collaboration across disciplines, cultures, and nations, the network can accelerate the development of the tools and frameworks needed to repair the past and restore the dead.

However, the success of this endeavor depends on more than just scientific and technological progress—it requires a commitment to ethical principles, inclusivity, and public engagement. Only by working together as a global community can we ensure that the future's efforts to repair the past are guided by wisdom, compassion, and justice.

Chapter 6: Ethical and Societal Implications

Section 1: Moral Justifications for Resurrection

Subsection 1: Fyodorov's Vision and the Beloved Community

The idea of universal scientific resurrection finds one of its earliest and most profound expressions in the work of Nikolai Fyodorovich Fyodorov, a 19th-century Russian philosopher, theologian, and visionary. Fyodorov's philosophy, often referred to as "The Philosophy of the Common Task," proposed a radical and deeply humanistic goal: the resurrection of all the dead through human effort, aided by science and technology. For Fyodorov, this was not merely a technical or scientific endeavor but a moral and spiritual imperative, rooted in the creation of a Beloved Community where humanity united in love, duty, and responsibility to overcome death itself.

This subsection explores Fyodorov's vision, his belief in the moral necessity of resurrection, and the role of the Beloved Community as a foundation for this universal task. It also examines how his ideas continue to inspire contemporary discussions about resurrection, transcending the boundaries of time, culture, and belief systems.

Nikolai Fyodorov and the Philosophy of the Common Task

Nikolai Fyodorov (1829–1903) was an enigmatic figure in Russian intellectual history, often considered the father of Russian Cosmism, a movement that combined scientific exploration with spiritual philosophy. Fyodorov's ideas were revolutionary in their scope, seeking to unite humanity in a shared effort to conquer death and restore life to all who had lived before.

1. The Core of Fyodorov's Vision

Fyodorov believed that death was the greatest injustice—a rupture in the continuity of life and a source of suffering for individuals and communities. Unlike most religious or philosophical systems that saw death as inevitable or natural,

Fyodorov argued that it was humanity's <u>moral duty</u> to overcome it.

- <u>Resurrection as a Common Task</u>: Fyodorov envisioned resurrection not as a divine miracle but as a collective human project. He believed that science and technology, guided by moral principles, could eventually restore life to the dead.
- <u>The Role of Science</u>: For Fyodorov, science was not a neutral or purely intellectual pursuit but a tool for achieving moral and spiritual goals. He saw the advancement of knowledge as intrinsically tied to humanity's responsibility to repair the injustices of the past.

2. A Universal and Inclusive Vision

Fyodorov's vision was universal in scope. He believed that resurrection should be extended to <u>all people, regardless of their time, place, or identity</u>. This inclusivity reflected his deep conviction that humanity is one interconnected family, and that the dead are as much a part of this family as the living.

- <u>Ancestral Responsibility</u>: Fyodorov emphasized the duty of the living to their ancestors. He saw resurrection as an act of gratitude and love, honoring the contributions of those who came before us.
- <u>Unity Across Generations</u>: In Fyodorov's view, resurrection would unite past, present, and future generations, creating a harmonious and eternal human community.

The Beloved Community: A Foundation for Resurrection

At the heart of Fyodorov's philosophy lies the concept of the <u>Beloved Community</u>, a vision of humanity united in love, cooperation, and shared purpose. Fyodorov believed that the fragmentation and conflicts of human society could be overcome through the pursuit of resurrection, which would inspire a new era of solidarity and moral growth.

1. The Ethical Imperative of Love

Fyodorov's vision of the Beloved Community was grounded in the principle of <u>love as a moral force</u>. He argued that love is not merely an emotion but a responsibility—a duty to care for others, including those who have died.

- Love as the Driving Force: Fyodorov saw love as the foundation of all ethical action. It is love for our ancestors, our fellow humans, and future generations that compels us to undertake the task of resurrection.
- Overcoming Division: The Beloved Community transcends divisions of race, nationality, class, and religion. In Fyodorov's view, resurrection unites all of humanity in a shared destiny, fostering a sense of universal belonging.

2. Cooperation and Collective Action

Resurrection, as conceived by Fyodorov, is not an individual endeavor but a collective task that requires the cooperation of all humanity. He believed that the pursuit of this goal would transform society, replacing competition and conflict with collaboration and mutual support.

- A New Model of Society: Fyodorov envisioned a future where the Beloved Community would prioritize the common good over individual interests. Science, technology, and culture would be directed toward the shared goal of overcoming death and restoring life.
- The End of Alienation: By uniting humanity in a common task, resurrection would heal the alienation and isolation that often characterize modern life. The Beloved Community would create a sense of purpose and connection that transcends individual existence.

3. Resurrection as Justice

For Fyodorov, resurrection was not only an act of love but also an act of justice. He saw death as a profound injustice that deprives individuals of their potential and separates families and communities. The Beloved Community, united in the pursuit of resurrection, would restore the balance of justice by repairing the wrongs of the past.

The Relevance of Fyodorov's Vision Today

Fyodorov's ideas, though rooted in the context of 19th-century Russia, continue to resonate in the 21st century. His visionary philosophy inspires contemporary discussions about the intersection of science, ethics, and the human condition.

1. Advances in Science and Technology
   Modern developments in fields such as quantum physics, synthetic biology, and artificial intelligence have brought humanity closer to realizing aspects of Fyodorov's vision. Technologies that once seemed like science fiction—genome editing, neural reconstruction, and quantum simulation—are now active areas of research, aligned with the goal of restoring life.

   - Scientific Resurrection: Fyodorov's belief in the potential of science to achieve resurrection aligns with current efforts to explore the boundaries of what is scientifically possible.
   - Global Collaboration: The interconnectedness of contemporary science reflects Fyodorov's vision of a unified humanity working together toward a common purpose.

2. Ethical and Philosophical Challenges
   The pursuit of resurrection raises profound ethical, philosophical, and spiritual questions that echo Fyodorov's concerns.

   - The Nature of Identity: Questions about what makes an individual "themselves"—their body, memories, or consciousness—are central to both Fyodorov's philosophy and modern debates about resurrection.
   - The Meaning of Death: Fyodorov's rejection of death as inevitable challenges traditional views and invites a reexamination of its role in human life and culture.

3. The Beloved Community in a Fragmented World
   In an era marked by division and conflict, Fyodorov's vision of the Beloved Community offers a powerful antidote. His emphasis

on love, cooperation, and shared purpose provides a model for addressing global challenges and fostering a sense of unity.

- A Moral Imperative: The idea of uniting humanity in a common task has never been more relevant, as the world faces existential threats such as climate change, inequality, and technological disruption.
- A Source of Hope: Fyodorov's vision reminds us that even the most ambitious goals—overcoming death and restoring life—are within reach when guided by love and collective effort.

Conclusion

Fyodorov's vision of universal scientific resurrection and the Beloved Community remains a source of inspiration and guidance for those who seek to repair the past and transcend the limits of human mortality. His philosophy challenges us to imagine a future where science and technology are harnessed not for personal gain or domination, but for the moral and spiritual task of uniting humanity in love and justice.

As we move closer to realizing the dream of resurrection, Fyodorov's insights remind us of the importance of grounding our efforts in compassion, responsibility, and shared purpose. The Beloved Community he envisioned offers not only a framework for resurrection but also a vision for a more harmonious and inclusive human future—one where the boundaries between life and death, past and present, are overcome by the enduring power of love.

Subsection 2: Expanding the Scope of Human Responsibility

The pursuit of universal scientific resurrection demands a profound reconsideration of what it means to be human and, more importantly, what responsibilities humanity bears—not only to itself but to the past, present, and future. Traditionally, human responsibility has been confined to the living and the immediate environment. However, the concept of resurrecting the dead expands this scope dramatically, placing new ethical,

- 123 -

philosophical, and practical obligations on humanity. It challenges us to transcend the limits of our existing moral frameworks and envision a future where our duties extend beyond the present moment, encompassing those who have come before us and those yet to come.

This subsection explores how resurrection compels humanity to expand its understanding of responsibility, addressing our obligations to the dead, the living, the environment, and the future. It also highlights how this expanded responsibility aligns with the vision of a more unified, compassionate, and forward-thinking human civilization.

Responsibility to the Dead

In most cultures, the dead are remembered and honored through rituals, memorials, and traditions. Yet, these acts are symbolic and limited to preserving memories, rather than restoring life. The idea of universal resurrection introduces a radical new dimension to our responsibility toward the dead: the obligation to repair the injustices of death and restore the lives of those who have passed.

1. Repairing the Injustice of Death
   Death, as Fyodorov and other thinkers have argued, is the ultimate injustice—an interruption of life and potential, and a source of grief for those left behind. By pursuing resurrection, humanity takes on the responsibility to undo the harm of death, treating it not as an immutable law but as a challenge to overcome.

   - Reclaiming Lost Potential: Every life contains unique potential—unrealized dreams, contributions to society, and relationships. Resurrection offers a way to restore this potential, allowing individuals to continue their journeys.
   - Honoring Ancestral Contributions: The achievements of past generations form the foundation of modern civilization. Resurrection acknowledges this debt, offering a way to repay it by restoring the lives of those who made these contributions.

2. Preserving the Memory of the Dead

The first step toward resurrection is the preservation and collection of information about the dead—their physical remains, genetic material, cultural records, and personal histories. This effort reflects a responsibility to ensure that no life is forgotten.

- A Universal Archive: Humanity must create systems to preserve the memories, artifacts, and genetic data of the dead, forming a comprehensive "archive of humanity."
- Equity in Remembrance: This responsibility must extend to all individuals, regardless of their social status, wealth, or historical prominence, ensuring that every life is valued equally.

Responsibility to the Living

While resurrection focuses on the dead, it also has profound implications for the living. The pursuit of this goal compels humanity to rethink its relationships, priorities, and collective responsibilities in the present.

1. Fostering Unity and Cooperation

The task of resurrection is so vast and complex that it requires global collaboration. This necessity fosters a sense of shared purpose, encouraging humanity to work together across borders, cultures, and disciplines.

- A Common Goal: Resurrection unites humanity in a shared endeavor, promoting cooperation and reducing divisions based on nationality, religion, or ideology.
- Building the Beloved Community: The act of working together on such a profound mission strengthens bonds between individuals and communities, creating a more compassionate and interconnected world.

2. Balancing Resources and Priorities

Resurrection efforts will require significant resources—scientific, financial, and social. Balancing these priorities with the needs of the living presents an ethical challenge, compelling humanity to act with care and wisdom.

- <u>Addressing Present Needs</u>: Efforts to repair the past must not come at the expense of addressing urgent issues in the present, such as poverty, inequality, and environmental degradation.

- <u>Creating Synergies</u>: Many of the technologies and systems developed for resurrection—such as advancements in healthcare, environmental restoration, and artificial intelligence—will also benefit the living, creating a virtuous cycle of progress.

<u>Responsibility to the Environment</u>

The environment plays a critical role in the process of resurrection, serving as both a source of information and a stage for the restored lives of the dead. Expanding the scope of human responsibility includes recognizing and addressing our obligations to the natural world.

1. <u>Environmental Restoration</u>

Resurrection requires an environment capable of supporting life. As such, the task compels humanity to prioritize the restoration and preservation of the planet's ecosystems.

- <u>Reversing Environmental Damage</u>: Decades of industrialization and exploitation have left scars on the Earth. Resurrection efforts must include initiatives to repair these damages, ensuring a habitable world for future generations and resurrected individuals alike.

- <u>Regenerating Biodiversity</u>: Just as death has claimed human lives, it has also led to the extinction of countless species. Resurrection technologies could be extended to restore biodiversity, bringing back lost species and repairing ecosystems.

2. <u>Sustainable Development</u>

The integration of resurrected individuals into society will require careful planning to ensure that resources are used sustainably and equitably.

- <u>Population and Resource Management</u>: Humanity must develop systems to balance the needs of the living, the resurrected, and the environment, ensuring that growth is harmonious and sustainable.

- Ethical Stewardship: As stewards of the planet, humanity bears the responsibility to create a world that can support life—not just for the present, but for all time.

Responsibility to the Future

The pursuit of resurrection is inherently future-oriented, requiring humanity to think not only about immediate outcomes but also about the long-term implications of its actions. This expanded sense of responsibility extends to future generations, ensuring that they inherit a world where the boundaries of life and death have been redefined.

1. Creating a Legacy of Hope

Resurrection represents a profound act of hope—a belief that the injustices of the past can be repaired and that humanity's potential is boundless.

- Inspiring Future Generations: The pursuit of resurrection can serve as a source of inspiration, encouraging future generations to dream boldly and work together to achieve seemingly impossible goals.
- A Vision of Progress: By expanding the scope of human responsibility, resurrection demonstrates that progress is not limited to technological advancements but includes moral and spiritual growth.

2. Ensuring Ethical Continuity

The technologies and systems developed for resurrection must be guided by ethical principles that ensure their responsible use in the future.

- Preventing Misuse: Humanity bears the responsibility to prevent the misuse of resurrection technologies, ensuring that they are not exploited for power, profit, or harm.

- Promoting Justice and Equity: Future generations must inherit a system that values all lives equally and prioritizes the common good over individual gain.

Expanding the Human Paradigm

The expanded scope of responsibility required for resurrection challenges humanity to rethink its role in the universe. It calls for a shift from a narrow, individualistic perspective to a broader, collective paradigm that embraces the interconnectedness of all life across time and space.

1. From Individualism to Collectivism
   Resurrection emphasizes the importance of collective action and shared purpose, fostering a sense of unity and mutual responsibility.

- A Shared Destiny: The task of resurrection reminds humanity that we are all part of a larger story, connected to the past and the future.
- Overcoming Isolation: By expanding our responsibilities, we transcend the isolation of individual existence and embrace a more holistic view of humanity.

2. From Mortality to Immortality
   Resurrection challenges the traditional understanding of mortality, offering a vision of life that is not bound by time.

- A New Relationship with Death: By addressing death as a problem to be solved, humanity redefines its relationship with mortality, transforming it from an end into a transition.
- Transcending Time: Responsibility no longer ends with the present generation—it extends across the entirety of human history, uniting past, present, and future in a seamless continuum.

Conclusion

The pursuit of universal scientific resurrection expands the scope of human responsibility in profound and transformative ways. It calls on humanity to take responsibility not only for the living but also for the dead, the environment, and future generations. This expanded responsibility reflects a new vision of what it means to be human—one that is rooted in compassion, unity, and a commitment to repairing the injustices of the past.

As humanity embarks on this ambitious journey, it must embrace the challenges and opportunities that come with this broader sense of duty. By doing so, we can create a future where life is no longer bound by the limitations of time and death, and where the bonds of love and responsibility unite all of humanity in a shared and eternal purpose.

Section 2: Addressing Potential Risks

Subsection 1: Risks of Unintended Consequences

The pursuit of universal scientific resurrection is one of humanity's most ambitious and morally complex endeavors. While it holds the promise of repairing the past and overcoming the boundaries of life and death, such an undertaking is fraught with risks of unintended consequences. These risks arise from the vast unknowns in the fields of physics, biology, and artificial intelligence that underpin the project, as well as the profound ethical, social, and environmental complexities it introduces.

As with any transformative technology, resurrection efforts could lead to outcomes that are unforeseen, mismanaged, or even harmful. This subsection explores the potential risks associated with universal resurrection, including technical failures, ethical dilemmas, societal disruptions, and environmental impacts. It argues for a cautious and responsible approach that prioritizes foresight, ethical oversight, and adaptability to navigate these challenges.

Technical Challenges and Failures

Resurrection relies on cutting-edge technologies in quantum mechanics, synthetic biology, and artificial intelligence, all of which are still in their infancy. The complexity and unpredictability of these systems can give rise to significant risks.

1. Incomplete or Inaccurate Reconstructions
The process of resurrecting individuals depends on retrieving and reconstructing vast amounts of data—genetic, neural, and quantum. Errors or gaps in this data could lead to incomplete or inaccurate resurrections, raising concerns about identity, authenticity, and human dignity.

- Loss of Data Integrity: If critical information about an individual's physical or mental state is missing or misinterpreted, the resulting reconstruction may not faithfully restore the original person.
- Unintended Variations: Small errors in genetic or neural reconstruction could lead to unintended changes in the individual's personality, memories, or physical appearance, creating ethical dilemmas about whether the resurrected person is still "themselves."

2. Unforeseen Interactions Between Technologies
Resurrection technologies involve the integration of multiple systems, such as quantum computing, bioprinting, and AI-driven neural simulations. These technologies may interact in unpredictable ways, leading to emergent behaviors or failures.

- Systemic Failures: A malfunction in one system could cascade into others, jeopardizing the entire reconstruction process.
- AI Misalignment: If artificial intelligence systems are not aligned with ethical guidelines, they could optimize for unintended outcomes, introducing biases or even harmful behaviors into the resurrection process.

3. Irreversible Mistakes
Resurrection experiments carry the risk of irreversible consequences, such as harm to the resurrected individual or damage to the original data archives. Once an error is made, it may be impossible to undo, particularly if the individual is already "restored" in an incomplete or distorted form.

Ethical and Philosophical Dilemmas

The resurrection of the dead raises profound ethical and philosophical questions, many of which have no clear answers. Without careful consideration, these dilemmas could lead to moral and societal crises.

1. Consent and Autonomy

The dead cannot provide consent to be resurrected, raising ethical questions about whether it is right to bring someone back without their explicit permission.

- Posthumous Autonomy: Does bringing someone back to life violate their autonomy, especially if they did not express a desire for resurrection during their lifetime?
- Proxy Decision-Making: Should family members, descendants, or society at large have the authority to decide who is resurrected and under what conditions?

2. Identity and Authenticity

Resurrection challenges the very notion of identity. If a person is reconstructed from incomplete data, or if their memories and personality are altered during the process, are they still the same person?

- The "Replica Problem": If the resurrected individual is not identical to the original, what moral status do they hold? Are they a continuation of the original person or a wholly new entity?
- Competing Claims of Identity: In cases where multiple reconstructions of the same individual are possible (e.g., based on different datasets), which version is the "real" one?

3. Inequity and Favoritism

The resurrection process may not be equally available to all, raising concerns about inequity and favoritism.

- Selective Resurrection: Economic, social, or political considerations could lead to the resurrection of some individuals while others are excluded, perpetuating existing inequalities.

- <u>Cultural Bias</u>: Historical figures from certain nations, cultures, or periods may be prioritized over others, leading to a skewed representation of the past.

4. <u>The Right to Die</u>
Some individuals may not wish to be resurrected, either for personal, religious, or philosophical reasons. Ignoring these preferences could violate their right to remain at peace.

<u>Societal Disruptions</u>

The resurrection of the dead would fundamentally alter the structure and dynamics of human society, introducing risks of conflict, instability, and unintended consequences.

1. <u>Overpopulation and Resource Strain</u>
The integration of resurrected individuals into society could lead to significant increases in population, placing immense strain on resources such as food, water, energy, and housing.

- <u>Competition for Resources</u>: As the resurrected re-enter the workforce, educational systems, and social services, competition for limited resources could exacerbate inequality and social tensions.
- <u>Environmental Impact</u>: A sudden population surge could accelerate environmental degradation, undermining efforts to create a sustainable future.

2. <u>Social and Cultural Fragmentation</u>
The resurrection of individuals from vastly different historical periods could introduce <u>cultural clashes</u> and disrupt societal cohesion.

- <u>Conflicting Values</u>: Individuals from different eras may hold conflicting beliefs, values, and worldviews, creating friction within communities.
- <u>Power Dynamics</u>: Resurrected individuals with significant historical influence (e.g., political leaders, inventors, or

philosophers) could disrupt modern power structures, leading to instability.

3. Psychological and Emotional Impact
The return of the dead could have profound psychological and emotional effects on the living.

- Family Dynamics: The resurrection of deceased loved ones could lead to complex emotional challenges, such as reconciling with the past or navigating changed relationships.
- Grief and Trauma: For some, the return of the dead might reopen old wounds or create new forms of grief and trauma, particularly if the resurrection process is imperfect.

Environmental Risks

The resurrection process itself could have unintended consequences for the environment, particularly if it involves large-scale resource use or technological interventions.

1. Energy Consumption
The technologies required for resurrection, such as quantum computing and bioprinting, are likely to be highly energy-intensive. Without careful planning, this could exacerbate global energy crises and contribute to climate change.

2. Impact on Ecosystems
The reintroduction of large numbers of individuals into the environment could disrupt ecosystems and biodiversity.

- Habitat Loss: Expanding infrastructure to accommodate the resurrected population may lead to habitat destruction.
- Extinction Risks: The resurrection of extinct species (a potential extension of the technology) could unbalance ecosystems, creating unforeseen consequences.

Mitigating the Risks of Unintended Consequences

To address these risks, humanity must adopt a cautious and deliberate approach to resurrection research and implementation. Key strategies include:

1. Ethical Oversight
Establish international ethical committees to guide resurrection efforts, ensuring that they align with principles of justice, equity, and respect for human dignity.

2. Scenario Planning and Risk Assessment
Use advanced modeling and simulation tools to anticipate potential unintended consequences, enabling proactive measures to minimize harm.

3. Global Governance
Create a global framework for regulating resurrection technologies, ensuring that they are used responsibly and equitably.

4. Public Engagement
Involve communities in discussions about resurrection, addressing concerns, and building consensus on how to proceed.

Conclusion

The risks of unintended consequences associated with universal scientific resurrection are significant and multi-faceted, encompassing technical, ethical, societal, and environmental challenges. While these risks should not deter humanity from pursuing this ambitious goal, they demand careful consideration and proactive management.

By recognizing and addressing these risks, humanity can navigate the complexities of resurrection responsibly, ensuring that the future's efforts to repair the past are guided by wisdom, compassion, and foresight. Only through a balanced approach can we achieve the promise of resurrection while preserving the dignity of the living, the dead, and the planet itself.

Subsection 2: Ensuring Equitable Access to Resurrection Technologies

The development of resurrection technologies represents a profound scientific and moral milestone, offering humanity the potential to bridge the gap between life and death by restoring those who have passed. However, with such transformative power comes the danger of inequality and exclusivity. If access to resurrection is limited by wealth, power, geography, or other forms of privilege, it risks becoming a tool for reinforcing existing disparities rather than a universal solution to the injustices of death.

Ensuring equitable access to resurrection technologies is not only a practical challenge but also an ethical imperative. This subsection explores the risks of inequality in resurrection efforts, examines the importance of creating fair and inclusive frameworks, and proposes strategies to ensure that this profound capability is available to all of humanity, regardless of status or circumstance.

The Risk of Inequitable Access

Without proper safeguards, resurrection technologies could become a tool of privilege, accessible only to the wealthy, powerful, or influential. The risks of inequitable access are multifaceted, encompassing economic, social, and cultural dimensions.

1. Economic Barriers
Resurrection technologies are likely to be resource-intensive, requiring significant investments in research, infrastructure, and materials. As a result, there is a risk that only those with substantial financial resources will be able to afford access.

- Privatized Resurrection: Companies or organizations that develop resurrection technologies may prioritize profit over equity, offering services exclusively to those who can pay exorbitant fees.
- Widening Inequality: If resurrection is limited to the wealthy, it could exacerbate existing inequalities, creating a world where the rich enjoy immortality while the poor are excluded from this transformative capability.

2. Geopolitical Disparities

Resurrection technologies might initially be developed in technologically advanced nations, creating a divide between countries that have access to these capabilities and those that do not.

- Global Inequality: Wealthy nations could monopolize resurrection technologies, leaving less developed nations behind and perpetuating global inequities.
- Technological Imperialism: The control of resurrection technologies by a few powerful nations or corporations could lead to exploitation or coercion of less powerful regions.

3. Cultural and Historical Biases

Decisions about who is resurrected—and how—may reflect the cultural, political, or historical biases of those in control of the technology.

- Selective Resurrection: Individuals from certain nations, cultures, or historical periods may be prioritized for resurrection, while others are neglected or excluded.
- Erasure of Marginalized Voices: Communities that have been historically marginalized or oppressed may face further exclusion if their ancestors and cultural histories are not given equal attention in resurrection efforts.

The Ethical Imperative of Equity

Equitable access to resurrection technologies is not merely a practical consideration—it is a moral obligation. Resurrection, by its very nature, seeks to address the injustices of death. To restrict access to this capability would contradict its foundational goal of universal justice and human dignity.

1. Resurrection as a Universal Right
The ability to restore life should not be treated as a privilege reserved for a select few but as a fundamental human right that belongs to all.

- Equality in Death and Life: Death affects all humans equally, regardless of their wealth, nationality, or status. Resurrection should reflect this universality by being accessible to all.
- Moral Responsibility: Those who develop resurrection technologies have a moral responsibility to ensure that their work benefits humanity as a whole, rather than serving narrow interests.

2. Honoring the Dignity of All Lives
Every human life has inherent value, and resurrection efforts must reflect this principle by treating all individuals with equal respect and importance.

- No Life Left Behind: Resurrection should aim to include everyone, regardless of their social, economic, or historical prominence. The laborer, the artist, and the ruler should be treated with equal care and consideration.
- Inclusion of Marginalized Histories: Efforts to reconstruct the past must prioritize the inclusion of voices and communities that have been historically erased or overlooked.

Designing Inclusive Frameworks for Access

To ensure equitable access to resurrection technologies, humanity must develop inclusive frameworks that prioritize justice, fairness, and universal benefit. These frameworks should address the economic, social, and cultural dimensions of equity.

1. Global Governance of Resurrection Technologies
Resurrection efforts should be overseen by an international body that represents the interests of all humanity, rather than being controlled by individual nations or corporations.

- A United Framework: An organization akin to the United Nations could be established to coordinate resurrection efforts, ensuring that all nations and communities have a voice in decision-making.
- Transparent Decision-Making: Policies and procedures for resurrection must be transparent, with clear criteria for how individuals are selected and prioritized.

2. Public Funding and Open Access
To prevent resurrection technologies from becoming privatized or exclusive, they should be funded and managed as a public good.

- Government and Nonprofit Funding: Resurrection research should be supported by governments, international organizations, and nonprofit institutions, ensuring that it is not driven solely by profit motives.
- Open-Source Technologies: Key resurrection technologies—such as quantum information retrieval systems, bioprinting tools, and neural reconstruction algorithms—should be developed as open-source platforms, accessible to researchers and practitioners worldwide.

3. Equitable Prioritization Criteria
In cases where resources are limited, clear and equitable criteria should guide the prioritization of who is resurrected.

- Historical Representation: Efforts should aim to include individuals from all historical periods, cultures, and regions, avoiding favoritism or bias.
- Focus on Underserved Communities: Priority could be given to individuals and communities that have been historically marginalized or whose contributions have been overlooked.

Building Public Trust and Engagement

Equitable access to resurrection technologies depends not only on institutional frameworks but also on public trust and involvement. Communities must feel that their voices are heard and that the process reflects their values and priorities.

1. Community Participation
Local and global communities should have a direct role in shaping resurrection efforts, ensuring that they reflect diverse perspectives and priorities.

- Public Forums and Dialogues: Open discussions about resurrection should be held in communities worldwide, inviting input from diverse stakeholders.
- Collaborative Decision-Making: Community representatives should be included in decision-making bodies, ensuring that local priorities are represented.

2. Education and Awareness
Public understanding of resurrection technologies is essential for building trust and ensuring informed participation.

- Accessible Information: Educational campaigns should provide clear, accessible explanations of resurrection technologies, their potential benefits, and their limitations.
- Addressing Concerns: Efforts should be made to address public concerns about equity, ethics, and unintended consequences, fostering an open and transparent dialogue.

A Vision of Universal Inclusivity

The ultimate goal of resurrection is to create a world where the past is repaired, the present is enriched, and the future is filled with hope. This vision can only be realized if resurrection technologies are developed and implemented in a way that reflects the principles of universal inclusivity.

1. A Shared Human Legacy

Resurrection is not just about restoring individuals—it is about preserving and honoring the shared legacy of humanity. Ensuring equitable access ensures that this legacy reflects the diversity and richness of human experience.

2. A Model for Global Justice

The equitable distribution of resurrection technologies can serve as a model for addressing other global challenges, demonstrating humanity's capacity to act with fairness, compassion, and collective purpose.

Conclusion

Ensuring equitable access to resurrection technologies is one of the most critical challenges of this transformative endeavor. Without careful planning and ethical oversight, resurrection risks deepening existing inequalities and creating new forms of exclusion.

By prioritizing transparency, inclusivity, and fairness, humanity can ensure that resurrection technologies fulfill their promise as a universal solution to the injustices of death. In doing so, we honor the value of every human life and take a significant step toward a future where the past, present, and future are united in a shared vision of justice and hope.

Part IV: Philosophical and Practical Challenges

Chapter 7: Grappling with Metaphysics and Identity

Section 1: The Nature of Self and Continuity

Subsection 1: What Does it Mean to Resurrect a Person?

The concept of resurrection has long been intertwined with human aspirations, appearing in religious, philosophical, and cultural narratives throughout history. However, the idea of scientific resurrection brings a new dimension to this ancient concept, shifting it from the realm of faith and myth into a potential technological and ethical reality. But before humanity can embark on such a monumental task, a fundamental question must be addressed: What does it actually mean to resurrect a person?

Resurrecting a person is not simply a matter of reanimating a body or restoring biological processes. It is a profoundly complex endeavor that involves questions of identity, continuity, and the nature of selfhood. This subsection explores these dimensions, examining the components of personal identity, the challenges of reconstructing a life, and the philosophical implications of "bringing someone back." By unpacking what it truly means to resurrect a person, we can better understand the scope and depth of this universal scientific task.

The Components of Personal Identity

Resurrecting a person requires more than restoring their physical body. A person's identity is a multi-faceted construct, shaped by biological, psychological, and social factors. Understanding these components is critical to defining what resurrection entails.

1. The Physical Body
   A person's body is often the most tangible aspect of their identity. It includes their unique genetic makeup, physical appearance, and biological functions.

   - Genetic Continuity: Resurrection would require the accurate reconstruction of an individual's DNA to recreate their physical form. This raises technical challenges, especially if genetic material is incomplete or degraded over time.

- Physical Appearance: Beyond genetics, attributes such as scars, birthmarks, and other physical characteristics are integral to how individuals are recognized by others and themselves.

2. The Mind and Consciousness

A person's mind—encompassing their memories, thoughts, personality, and consciousness—is central to their identity. Resurrecting a person must account for these intangible but essential aspects.

- Memories: Memories are the threads that connect a person to their past, shaping their sense of self. Resurrection would require the retrieval and reconstruction of memories, potentially from neural patterns, historical records, or other data sources.
- Personality: Traits such as temperament, preferences, and habits define how a person interacts with the world. These aspects must also be faithfully restored for the resurrected individual to feel authentic.
- Consciousness: The most elusive aspect of identity, consciousness raises profound questions. Can it be fully reconstructed, or is it inherently tied to the original biological processes of the brain?

3. Social and Relational Identity

A person's identity is not formed in isolation; it is deeply influenced by their relationships, social roles, and cultural context.

- Relationships: Family, friendships, and social bonds play a crucial role in shaping who a person is. Resurrection must consider how these relationships are re-established or redefined.
- Cultural and Historical Context: A person's identity is rooted in the time and place they lived. Resurrecting someone without accounting for their cultural and historical context risks severing them from the world they knew.

Reconstructing a Life

Once the components of identity are understood, the next challenge lies in reconstructing them. Resurrection is not merely about recreating isolated elements but about <u>reassembling a cohesive and continuous self</u>. This task raises significant technical and philosophical questions.

1. <u>The Continuity of Identity</u>
   For a resurrection to be meaningful, the resurrected individual must feel like a continuation of their former self. This continuity depends on preserving key aspects of their identity.

- <u>Temporal Gaps</u>: If centuries or millennia have passed since a person's death, how can their identity remain continuous? Resurrection must bridge this temporal gap, ensuring that the individual's sense of self is intact.
- <u>Integration of Restored Memories</u>: Memories retrieved from various sources must be integrated seamlessly into the resurrected individual's mind. Any inconsistencies or gaps could disrupt their sense of self.

2. <u>The Role of Technology</u>
   Advanced technologies will play a central role in reconstructing a person's body, mind, and identity.

- <u>Quantum Information Retrieval</u>: Theoretical approaches, such as using quantum traces or physical remnants to retrieve information about the individual, may allow for the reconstruction of both physical and mental aspects.
- <u>AI-Assisted Reconstruction</u>: Artificial intelligence could assist in piecing together fragmented data, simulating neural patterns, and reconstructing memories. However, this raises questions about accuracy and authenticity.
- <u>Biological Engineering</u>: Technologies such as bioprinting and stem cell regeneration could recreate the person's physical body, ensuring biological continuity.

3. Ethical and Practical Challenges

The reconstruction of a person's identity is fraught with challenges, both ethical and practical.

- Incomplete Data: In many cases, the available data about a person may be incomplete, requiring scientists to make assumptions or extrapolations. This raises questions about whether the reconstructed person is truly "them."
- Consent and Authenticity: Without the ability to obtain consent from the deceased, how can we be certain that they would want to be resurrected? Additionally, if significant parts of their identity are reconstructed using artificial means, does their authenticity come into question?

Philosophical Implications of Resurrection

The act of resurrecting a person is not only a technical achievement but also a profound philosophical statement about the nature of life, death, and identity. It forces humanity to confront some of its deepest questions.

1. What Makes a Person "Themselves"?

Is a person defined by their physical body, their memories, their consciousness, or some combination of these? Resurrection tests the boundaries of personal identity, challenging us to define what truly makes someone who they are.

- The Ship of Theseus Problem: If a person's body, mind, and memories are reconstructed piece by piece, is the resulting individual still the same person, or are they a new entity?
- The Role of Memory: If memories are central to identity, how do we address gaps, inconsistencies, or artificially reconstructed memories?

2. The Nature of Life and Death

Resurrection redefines the boundaries between life and death, raising questions about what it means to truly live.

- Reversing Death: If death is no longer permanent, how does this change our understanding of mortality and the human experience?
- The Ethics of Immortality: Does resurrection imply a form of immortality, and if so, is this desirable or ethical?

3. The Relationship Between the Past and the Present
Resurrecting individuals from the past creates a unique relationship between the present and history.

- Integration into the Modern World: How do resurrected individuals adapt to a world that may be vastly different from the one they knew?
- Reconciliation with History: Resurrection offers an opportunity to repair the injustices of the past, but it also raises questions about how much of the past should be restored.

Conclusion

To resurrect a person is to undertake a task of profound complexity, involving the reconstruction of their body, mind, and identity. It is not merely the restoration of life but the restoration of selfhood, with all its intricacies and nuances.

By exploring the components of identity, the challenges of reconstruction, and the philosophical implications of resurrection, we can begin to understand the true scope of this universal scientific endeavor. Resurrection is not just a technical problem to be solved but a deeply human project that forces us to confront what it means to be alive, to be ourselves, and to exist in continuity with the past. It is in answering these questions that humanity may find not only the keys to resurrecting the dead but also a deeper understanding of life itself.

Subsection 2: Memory, Identity, and the Soul in Resurrection

At the heart of any attempt to resurrect a person lies a profound question: What makes a person who they are? Is it their memories, their personality, their physical body, or something more ineffable, such as their soul? The interplay between memory, identity, and the soul is both a scientific and philosophical puzzle, one that must be addressed to make universal scientific resurrection meaningful and authentic.

This subsection explores the role of memory in shaping identity, the challenges of preserving and restoring a coherent sense of self, and the question of whether the soul—however it is defined—plays an essential role in resurrection. It examines these themes from scientific, philosophical, and ethical perspectives, offering insights into how they influence the feasibility and desirability of universal resurrection.

The Role of Memory in Identity

Memory is often seen as the cornerstone of personal identity. It connects us to our past, informs our present, and shapes our future. When considering resurrection, the accurate restoration of an individual's memories is essential to preserving their sense of self.

1. Memory as the Foundation of Selfhood
Our memories are more than just a record of past events; they are the threads that weave together our identity. Without memory, a person might lose their sense of continuity and self-recognition.

- Autobiographical Memory: This includes memories of personal experiences, such as childhood events, relationships, and life milestones. These memories provide the narrative structure of a person's life.
- Implicit and Emotional Memory: Beyond conscious recollections, implicit memories and emotional associations contribute to personality, behavior, and preferences.
- Cognitive Skills and Knowledge: Memories of acquired skills, such as language, problem-solving, and cultural knowledge, also define who we are.

2. Challenges in Restoring Memory
The process of accurately restoring memories in resurrection presents significant challenges, both technical and philosophical.

- Incomplete Data: Memories may be stored in the physical structure of the brain, but they are fragile and subject to decay after death. Retrieving and reconstructing lost memories, especially when no direct record exists, may be immensely difficult.
- Reconstruction vs. Authenticity: If memories are reconstructed from external sources, such as historical records or third-party accounts, how can we ensure they are authentic and not distorted? Would the individual still recognize themselves in these memories?
- Memory Gaps: How should gaps in memory be handled? Would resurrected individuals feel incomplete or disoriented if significant parts of their past are missing?

3. Memory and the Present Self
Crucially, memory is not static—it is constantly reshaped by our experiences and perceptions. A resurrected individual may reinterpret or even reject certain memories based on their new context, raising questions about whether resurrection restores the person they were or creates someone new.

Identity: Beyond Memory and the Body

While memory is central to identity, it is not the only factor. A person's identity also encompasses their personality, relationships, and their sense of self-awareness. Resurrection must address these additional dimensions to fully restore an individual.

1. Personality and Behavioral Continuity
Personality traits—such as temperament, values, and habits—are crucial to identity. While some of these traits may be shaped by genetics, others are influenced by life experiences and social interactions.

- Stability vs. Change: Personality is not fixed; it evolves over time. Resurrection must decide which version of a person's personality to restore—should it reflect their state at the time of death, or an earlier, perhaps "idealized" version of themselves?

- Behavioral Reconstruction: Beyond memories, resurrection may involve simulating neural patterns that govern behavior. However, this introduces ethical concerns about whether such simulations can fully capture the complexity of human personality.

2. Social Identity and Relationships

A person's identity is deeply tied to their relationships and social roles. Resurrection must consider how these dynamics are re-established or altered.

- Relational Identity: How does the resurrected individual fit into a world where their loved ones may have also changed or passed on?
- Cultural and Historical Context: A person's sense of self is shaped by the culture and time they lived in. Resurrection must address whether individuals can adapt to a modern world that may be vastly different from their original context.

3. The Question of Authenticity

If an individual's memories, personality, and relationships are reconstructed using advanced technologies, does this create a genuine continuation of the original person, or a new entity altogether? This question lies at the heart of debates about personal identity in resurrection.

- The Replica Problem: Is a resurrected individual a true continuation of their former self, or merely a highly accurate replica?
- Continuity of Consciousness: Without a continuous stream of consciousness from life to resurrection, can the resurrected individual claim to be the same person?

The Soul: A Scientific and Philosophical Perspective

The concept of the soul complicates the scientific discourse on resurrection, as it introduces questions that transcend materialism. While science often focuses on the physical and cognitive aspects of identity, many philosophical and religious traditions assert that the soul is the essence of a person. How might resurrection account for this dimension?

1. Defining the Soul

The soul is understood in various ways across cultures and traditions, ranging from a divine and eternal essence to a metaphor for consciousness and individuality.

- Dualistic Views: In many religious traditions, the soul is seen as separate from the body and mind, surviving death and persisting independently. Resurrection, in this view, would involve reuniting the soul with a reconstructed body.
- Materialist Interpretations: From a scientific perspective, the "soul" might be interpreted as the emergent properties of consciousness, arising from the brain's physical structure and neural activity.

2. Can the Soul Be Resurrected?

If the soul is immaterial, how can science address its role in resurrection? Several possibilities emerge:

- Soul Retrieval: Some theories posit that the soul might exist as information or energy that could be retrieved and reintegrated into a reconstructed body. However, this remains speculative and unproven.
- Emergent Soul: If the soul is seen as an emergent property of the brain, it might naturally reappear during the reconstruction of neural patterns and consciousness. In this view, the resurrected individual's soul would be a continuation of their identity, inseparable from their reconstructed self.

3. Ethical and Spiritual Implications

The inclusion of the soul in resurrection raises profound ethical and spiritual questions.

- Respect for Religious Beliefs: Resurrection efforts must respect diverse beliefs about the soul, ensuring that individuals and communities feel their spiritual values are honored.

- The Mystery of the Soul: Even if science cannot definitively address the existence or nature of the soul, its mystery may inspire a sense of humility and reverence in the pursuit of resurrection.

Reconciling Memory, Identity, and the Soul

The interplay between memory, identity, and the soul presents both challenges and opportunities for resurrection. To reconcile these dimensions, resurrection efforts must adopt a holistic approach that integrates scientific rigor with philosophical and ethical sensitivity.

1. A Multi-Dimensional View of Identity

Resurrection must move beyond a narrow focus on physical or cognitive reconstruction to embrace a multi-dimensional view of identity that includes memory, personality, relationships, and potentially the soul.

2. Ethical and Cultural Sensitivity

Efforts to restore individuals must respect their cultural, spiritual, and personal beliefs, ensuring that the process aligns with their values and desires.

3. The Imperfect but Noble Task

While perfect fidelity in resurrection may be unattainable, the pursuit of this goal reflects humanity's deepest aspirations to repair the past, honor the dead, and transcend the boundaries of mortality.

Conclusion

Memory, identity, and the soul are inseparable from the question of what it means to resurrect a person. Memory provides the narrative of a life, identity encompasses the continuity of self, and the soul—whether material or immaterial—touches on the essence of being. Together, these dimensions shape the profound complexity of resurrection, requiring humanity to confront its deepest questions about life, death, and what it means to truly exist.

As we explore the possibilities of universal scientific resurrection, we must strive to honor these dimensions with care and humility, ensuring that the process not only restores individuals but also preserves the integrity of their humanity. In doing so, we take a step closer to bridging the gap between the past and the future, guided by the enduring mystery of what it means to be alive.

Section 2: Reconciling Different Philosophical Frameworks

Subsection 1: Comparing Fyodorov, Kant, and Whitehead

The idea of universal scientific resurrection invites us to engage with some of the most profound philosophical questions ever asked: What is humanity's ultimate purpose? How do we reconcile the past with the future? And what role does human agency play in reshaping the world? To explore these questions, it is helpful to turn to the ideas of three towering thinkers: Nikolai Fyodorov, Immanuel Kant, and Alfred North Whitehead.

Though separated by time, discipline, and cultural context, Fyodorov, Kant, and Whitehead each offer insights that resonate with the ambitions and challenges of resurrection. Fyodorov's vision of cosmic unity through resurrection, Kant's emphasis on moral duty and the limits of reason, and Whitehead's focus on process and interconnectedness provide complementary frameworks for understanding the philosophical foundations of this audacious endeavor. This subsection compares their ideas, highlighting areas of convergence, divergence, and relevance to the concept of universal resurrection.

Nikolai Fyodorov: The Philosophy of the Common Task

Nikolai Fyodorov (1829–1903), a Russian philosopher and visionary, is often regarded as the father of the idea of universal scientific resurrection. His philosophy, known as the Common Task, posits that humanity's highest moral duty is to overcome death and restore life to all who have ever lived.

1. Resurrection as Humanity's Moral Obligation

Fyodorov viewed death as the greatest injustice and believed that humanity, through collective scientific effort, could and must overcome it. For him, resurrection was not merely a religious or symbolic act but a practical task.

- Unity and Brotherhood: Fyodorov emphasized the importance of collective action, arguing that humanity must unite to fulfill the shared goal of resurrecting the dead and creating a harmonious cosmos.
- Immortality and Responsibility: He saw immortality not as an individual pursuit but as a responsibility to the generations that came before us. Resurrection, in his view, was both an act of justice and a means of creating universal equality.

2. Technological Optimism and Cosmic Vision

Fyodorov's vision extended beyond Earth to encompass the entire cosmos. He believed that humanity's mastery of science and technology would eventually enable the resurrection of all beings, even reconstructing those whose remains had been scattered across space and time.

- Scientific Resurrection: Fyodorov anticipated the development of technologies that could retrieve and reconstruct the physical and informational essence of the dead.
- Cosmic Harmony: For Fyodorov, resurrection was not just about restoring life but about achieving a harmonious relationship between humanity, nature, and the cosmos.

3. Tensions and Critiques

While Fyodorov's vision is inspiring, critics have noted its utopian and speculative nature. His reliance on future technologies and his assumption of universal cooperation raise practical and ethical questions about feasibility and inclusivity.

Immanuel Kant: Duty, Reason, and the Limits of Knowledge

Immanuel Kant (1724–1804), a German philosopher and one of the central figures of modern philosophy, did not explicitly address resurrection in his works. However, his ideas about moral duty, human reason, and the nature of the self provide a critical lens for evaluating the philosophical underpinnings of resurrection.

1. The Categorical Imperative and Moral Duty
Kant's ethical framework, based on the categorical imperative, emphasizes the universality of moral principles. His philosophy suggests that moral actions must respect the dignity and autonomy of all individuals.

- Human Dignity: Kant's insistence on treating every person as an end in themselves aligns with the ethical dimension of resurrection, particularly its aim to honor and restore the intrinsic value of every human life.
- Universal Duty: While Kant focused on moral duties in the present, his framework could be extended to argue that humanity has a duty to repair past injustices, including the finality of death.

2. Reason and the Limits of Knowledge
Kant famously argued that human reason is limited in its ability to comprehend metaphysical questions, such as the immortality of the soul or the existence of God.

- Faith and Practical Reason: While Kant believed that questions of immortality could not be proven through reason, he maintained that they were necessary postulates of practical reason, essential for grounding moral action.
- Skepticism Toward Speculation: Kant's cautious approach contrasts with Fyodorov's speculative optimism. His philosophy reminds us to balance visionary goals with humility about the limits of human knowledge.

3. Temporal and Ethical Concerns
Kant's emphasis on autonomy raises ethical questions about resurrection. For example, would resurrecting someone without their consent violate their dignity? His philosophy also challenges us to consider the implications of resurrecting individuals from a vastly different historical and cultural context.

Alfred North Whitehead: Process, Interconnectedness, and Creativity

Alfred North Whitehead (1861–1947), an English mathematician and philosopher, is best known for his philosophy of <u>process</u> and his emphasis on the dynamic, interconnected nature of reality. Whitehead's ideas offer a unique perspective on resurrection, particularly in their focus on relationships, change, and creativity.

1. Process Philosophy and the Fluidity of Identity

Whitehead rejected static notions of identity and instead viewed reality as a process of becoming. For him, individuals are defined not by fixed essences but by their relationships and experiences.

- <u>Relational Identity</u>: Whitehead's philosophy suggests that resurrecting a person would involve not only restoring their physical and mental attributes but also recreating the web of relationships that shaped their existence.
- <u>Dynamic Continuity</u>: His emphasis on change raises questions about whether a resurrected person, adapted to a new context, could still maintain continuity with their former self.

2. Interconnectedness and the Web of Life

Whitehead's vision of reality as an interconnected whole aligns with Fyodorov's emphasis on universal harmony. Resurrection, in this view, is not an isolated act but part of a broader project to repair and enhance the interconnected fabric of existence.

- <u>Cosmic Creativity</u>: Whitehead's concept of creativity as the driving force of reality resonates with the idea of resurrection as a creative act that transforms the past and opens new possibilities for the future.
- <u>Ecological Implications</u>: Whitehead's focus on the environment and relationships suggests that resurrection must account for its impact on the broader ecological and cosmic system.

3. Ethical and Philosophical Implications

Whitehead's process philosophy challenges static ethical frameworks, emphasizing the importance of adaptability and creativity in addressing new challenges. His ideas encourage us to view resurrection not as a return to the past but as a forward-looking act of renewal.

Points of Convergence and Divergence

Though Fyodorov, Kant, and Whitehead approached their philosophies from different angles, their ideas intersect in ways that illuminate the challenges and possibilities of universal scientific resurrection.

1. Shared Emphasis on Universal Values
- Fyodorov's call for collective action, Kant's universal ethics, and Whitehead's relational philosophy all highlight the importance of transcending individualism to embrace a shared vision of humanity.
- All three thinkers grapple with the tension between individual autonomy and collective responsibility, a key issue in resurrection efforts.

2. Divergent Views on Speculation and Feasibility
- Fyodorov's speculative optimism contrasts with Kant's skepticism about the limits of reason. Whitehead, with his emphasis on creativity and process, provides a middle ground, emphasizing the potential for transformation without presuming certainty.

3. Complementary Ethical Frameworks
- Kant provides a rigorous ethical foundation for considering the dignity and autonomy of the resurrected.
- Fyodorov offers a moral imperative to repair the past and overcome death.
- Whitehead invites us to think dynamically and relationally, ensuring that resurrection aligns with the broader web of life.

Conclusion

Comparing Fyodorov, Kant, and Whitehead reveals the richness and complexity of the philosophical foundations underlying universal scientific resurrection. Fyodorov's visionary optimism, Kant's ethical rigor, and Whitehead's relational dynamism offer complementary perspectives that deepen our understanding of what it means to repair the past and create a harmonious future.

By synthesizing these ideas, we can approach resurrection not merely as a technical challenge but as a profound moral and philosophical undertaking—one that requires us to balance visionary ambition with ethical humility and creative adaptability. In this way, the wisdom of these three thinkers can guide humanity as it seeks to transcend the boundaries of life and death in the pursuit of justice, compassion, and universal unity.

Subsection 2: Bridging Eastern and Western Philosophical Traditions

The concept of universal scientific resurrection is not confined to any single cultural or philosophical tradition. It resonates across the vast spectrum of human thought, drawing upon both Eastern and Western philosophies to address fundamental questions about life, death, identity, and the possibility of restoration. While these traditions differ in their approaches to metaphysics, ethics, and the nature of existence, they also complement one another in ways that enrich our understanding of what it means to repair the past and resurrect the dead.

This subsection explores the intersections and contrasts between Eastern and Western philosophical traditions as they relate to resurrection. By bridging these traditions, we can uncover a more holistic and inclusive framework for approaching the profound challenges and opportunities of universal resurrection.

Western Philosophical Traditions: Individualism, Rationality, and the Ethics of Restoration

Western philosophy often emphasizes the individual self, the power of reason, and the ethical dimensions of human action. These themes have shaped Western approaches to questions of mortality and the possibility of resurrection.

1. Individualism and the Self
- In many Western traditions, particularly those influenced by Cartesian dualism, the self is understood as a distinct, autonomous entity. This view foregrounds the importance of preserving individual identity in any resurrection effort.
- Philosophers like John Locke have explored continuity of selfhood through memory, suggesting that personal identity is tied to the continuity of consciousness. This view aligns with the idea that resurrecting an individual requires the restoration of their unique memories and personality.

2. Reason and Mastery Over Nature
- Western thought, particularly in its modern period, has often celebrated the power of human reason to understand and transform the natural world. This intellectual tradition underpins the scientific optimism behind resurrection technologies.
- Thinkers like Francis Bacon and René Descartes envisioned science as a tool for mastering nature and overcoming human limitations, including death. The concept of resurrection as a scientific endeavor reflects this legacy.

3. Ethical Responsibility and Moral Duty
- Western philosophy's emphasis on ethics, particularly in the works of Immanuel Kant, frames resurrection as a moral obligation. Kant's categorical imperative—treating every individual as an end in themselves—resonates with the goal of honoring and restoring the dignity of all who have lived.
- The Judeo-Christian tradition, deeply embedded in Western thought, offers the theological concept of resurrection as an act of divine justice and salvation. While universal scientific resurrection

departs from theology, it echoes the ethical impulse to repair the injustices of death.

Eastern Philosophical Traditions: Interconnectedness, Impermanence, and the Continuum of Life

Eastern philosophies, by contrast, often emphasize interconnectedness, the impermanence of individual identity, and the cyclical nature of existence. These ideas provide a complementary perspective that challenges and enriches the Western focus on individuality.

1. Interconnectedness and the Collective Self
- In many Eastern traditions, such as Buddhism and Hinduism, the self is not seen as a fixed, autonomous entity but as part of a larger web of interdependent relationships.
- Dependent Origination (Buddhism) teaches that all phenomena, including human identity, arise from a network of causes and conditions. Resurrection, from this perspective, would not only involve reconstructing an individual but also restoring the relationships and contexts that define their existence.

2. Impermanence and the Fluid Nature of Identity
- The concept of Anicca (impermanence) in Buddhism challenges the notion of a permanent self. It suggests that identity is constantly evolving, shaped by experiences and circumstances.
- In Taoism, the self is understood as part of the dynamic flow of the Tao, the fundamental principle of the universe. Resurrection, in this context, could be seen as a reentry into the ever-changing flow of existence, rather than a return to a fixed state.

3. Cyclicality and the Continuum of Life
- Eastern traditions often emphasize the cyclical nature of life and death. In Hinduism, the concept of Samsara (the cycle of birth, death, and rebirth) suggests that life and death are part of an ongoing process of transformation.
- Resurrection, viewed through this lens, might not be about restoring a static past but about continuing the journey of existence

in a new form. It invites reflection on how resurrected individuals would integrate into the ongoing cycle of life.

Points of Intersection: A Holistic View of Resurrection

Despite their differences, Eastern and Western traditions converge on several key ideas that can inform a more holistic approach to resurrection.

1. The Value of Life
- Both traditions affirm the intrinsic value of life, whether viewed through the Western lens of individual dignity or the Eastern lens of interconnectedness. Resurrection, in honoring and restoring life, aligns with this shared value.

2. Repairing the Past
- Western ethics emphasizes the moral responsibility to repair past injustices, while Eastern philosophies highlight the continuity of actions and their consequences (e.g., karma). Together, these perspectives underscore the importance of resurrection as an act of justice and reconciliation.

3. Transcending Mortality
- Western traditions often seek to overcome death through the assertion of human reason and technological mastery, while Eastern traditions view death as a transformation within the larger continuum of existence. Resurrection, as a synthesis of these views, can be understood as both a restoration and a transformation.

Points of Tension: Reconciling Individualism and Interconnectedness

While Eastern and Western philosophies offer complementary insights, they also present tensions that must be addressed in the context of resurrection.

1. The Nature of the Self
   - Western traditions prioritize the restoration of the individual self, while Eastern traditions emphasize the fluid, interconnected nature of identity. Resurrection efforts must navigate this tension by balancing the preservation of individual identity with the recognition of its relational and dynamic aspects.

2. The Purpose of Resurrection
   - Western frameworks often view resurrection as a means of achieving justice and immortality, while Eastern frameworks might question whether such goals align with the natural flow of existence. This raises ethical questions about whether resurrection disrupts or harmonizes with the broader patterns of life and death.

3. Consent and Autonomy
   - Western ethics emphasizes individual consent and autonomy, raising questions about whether resurrected individuals would want to return to life. Eastern traditions, with their focus on interdependence, might place greater emphasis on the collective good and the interconnected consequences of resurrection.

Bridging the Traditions: Toward an Inclusive Philosophy of Resurrection

To create a universal framework for resurrection, it is essential to integrate the strengths of both Eastern and Western traditions, forging a philosophy that honors their differences while embracing their commonalities.

1. Embracing Pluralism
   - Resurrection must be approached with humility and openness, recognizing that no single tradition holds all the answers. By drawing from diverse philosophical insights, we can create a more inclusive and nuanced vision.

2. Balancing Individuality and Interconnectedness
   - Efforts to resurrect individuals must respect their unique identities while acknowledging the relational and contextual

aspects of selfhood. This balance reflects the interplay between Western individualism and Eastern interdependence.

3. Honoring Life in All Its Forms
- Whether viewed as a restoration of individual dignity or a continuation within the web of existence, resurrection must honor the value of life in all its dimensions—biological, psychological, social, and spiritual.

4. Resurrection as Transformation
- Bridging the traditions invites us to view resurrection not merely as a return to the past but as a transformative act that integrates the old with the new, creating possibilities for growth, renewal, and harmony.

Conclusion

Bridging Eastern and Western philosophical traditions provides a richer and more comprehensive framework for understanding universal scientific resurrection. Western thought contributes its focus on individual dignity, ethical responsibility, and technological mastery, while Eastern thought offers insights into interconnectedness, impermanence, and the cyclicality of life.

By synthesizing these perspectives, we can approach resurrection not merely as a technical challenge but as a profound act of reconciliation—one that honors the diversity of human thought and strives to repair the past while embracing the possibilities of the future. In doing so, we move closer to a vision of resurrection that is not only universal in its scope but also inclusive in its spirit.

Chapter 8: Overcoming Scientific and Technical Barriers

Section 1: Current Gaps in Knowledge

Subsection 1: Quantum Mechanics and the Mystery of Time

The concept of <u>universal scientific resurrection</u> inherently challenges conventional understandings of time and causality. To resurrect the dead, one must grapple with the past—retrieving information, reconstructing identities, and bridging the temporal gap between what was and what is. Such an endeavor naturally draws our attention to <u>quantum mechanics,</u> a field of science that has profoundly reshaped our understanding of time and the fundamental nature of reality.

Quantum mechanics, with its probabilistic nature and counterintuitive phenomena, offers insights into the malleability of time and the interconnectedness of events across the past, present, and future. By exploring quantum theories such as superposition, entanglement, and the arrow of time, we can begin to understand the profound implications for scientific resurrection. This subsection delves into the relationship between quantum mechanics and time, examining how the mysteries of quantum phenomena might one day enable humanity to repair the past.

<u>The Quantum Nature of Reality</u>

Quantum mechanics reveals a reality that is far more complex and dynamic than the deterministic framework of classical physics. At the heart of this quantum reality lie phenomena that challenge our intuitive understanding of time and causality.

1. <u>Superposition: Coexistence of States</u>
 - In the quantum world, particles can exist in multiple states simultaneously, a phenomenon known as <u>superposition.</u> For example, an electron can be in two places at once or have multiple energy levels until it is observed.
 - This suggests that the past may not be as fixed as it seems. If all possible states of a system coexist, then the information

required to reconstruct the past may still exist in a quantum superposition, waiting to be accessed or disentangled.

2. Quantum Entanglement: The Interconnectedness of Events

- **Entanglement** occurs when two particles become linked in such a way that the state of one instantly influences the state of the other, regardless of the distance between them. This "spooky action at a distance," as Einstein called it, implies that events in the universe are deeply interconnected.

- Entanglement raises intriguing possibilities for resurrection. Could the information about a person—encoded in the quantum states of particles—be retrieved through the entangled relationships that persist across time and space?

3. The Observer Effect and the Role of Consciousness

- In quantum mechanics, the act of observation influences the state of a system. This raises profound questions about the role of consciousness in shaping reality.

- If observation collapses a quantum state into a definite outcome, might it be possible to "observe" or reconstruct the quantum information of the past, effectively retrieving the essence of a deceased individual?

The Arrow of Time: Irreversibility and Reversibility

One of the central mysteries of time is its apparent irreversibility. While the laws of physics are largely time-symmetric—meaning they work the same forward and backward in time—our everyday experience of time flows in one direction, from past to future. This phenomenon is known as the arrow of time.

1. The Thermodynamic Arrow of Time

- The second law of thermodynamics states that entropy, or disorder, in a closed system tends to increase over time. This gives rise to the forward flow of time as we perceive it.

- Resurrection challenges the thermodynamic arrow of time by seeking to reverse processes that have led to decay and death. To achieve this, science must find ways to locally decrease entropy without violating fundamental physical laws.

2. Quantum Time Reversal

- While classical processes are irreversible, quantum systems exhibit the possibility of time reversal under certain conditions. Experiments have demonstrated that quantum states can be "rewound" to earlier configurations, effectively undoing changes.
- This raises the tantalizing possibility of using quantum processes to reverse the effects of time on a person's physical and informational essence, enabling their resurrection.

3. The Role of Information in Time

- In the quantum view, information is never truly lost, even in processes like the evaporation of black holes (as suggested by the Hawking information paradox). This principle suggests that the information about a person's physical and mental state may be preserved at a fundamental level, even after death.
- If this information exists, resurrection could involve retrieving and reconstructing it, effectively "rewinding" the individual's quantum state to a prior configuration.

Quantum Theories of Time and Their Implications

Quantum mechanics has given rise to various theories of time, each with profound implications for the possibility of resurrection. These theories challenge the traditional view of time as a linear sequence of events and open up new ways of thinking about the relationship between the past, present, and future.

1. The Block Universe: Time as a Static Dimension

- According to the block universe theory, time is not a flowing river but a static dimension in which the past, present, and future all coexist. This view aligns with Einstein's theory of relativity, which treats time as a fourth dimension of spacetime.
- If the block universe is correct, the past is never truly "lost" but continues to exist as part of the spacetime fabric. Resurrection, in this context, might involve accessing and retrieving the information encoded in the past sections of the block universe.

2. The Many-Worlds Interpretation: Parallel Timelines

- The many-worlds interpretation of quantum mechanics suggests that every quantum event spawns multiple parallel universes, each representing a different outcome.
- This raises the possibility that versions of deceased individuals might still exist in alternate timelines. Resurrection could involve accessing these parallel universes to retrieve or recreate lost individuals.

3. Quantum Loop Gravity and Discrete Time

- In some quantum theories, such as loop quantum gravity, time is not continuous but consists of discrete units. This granular view of time suggests that the past could be reconstructed by piecing together these quantum "time loops."
- Resurrection might involve navigating these discrete units of time to reconstruct the sequence of events and restore a person's identity.

Quantum Resurrection: Speculative Applications

The intersection of quantum mechanics and resurrection remains speculative but offers intriguing possibilities for future exploration.

1. Quantum Information Retrieval

- Techniques such as quantum tomography could be used to reconstruct the quantum state of a person based on traces of information left behind in the environment.
- Advanced quantum computers might one day simulate the complex interactions that define a person's identity, effectively "resurrecting" their quantum essence.

2. Entanglement-Based Resurrection

- If the particles that once composed a person remain entangled with the universe, it may be possible to trace and reconstruct their quantum state. This would require unprecedented advances in quantum technology and a deep understanding of entanglement mechanics.

3. The Role of Quantum Consciousness

- Some theories, such as Orchestrated Objective Reduction (proposed by Roger Penrose and Stuart Hameroff), suggest that consciousness itself may have a quantum basis. If true, resurrection might involve restoring not only the physical body and memories but also the quantum processes underlying consciousness.

Philosophical and Ethical Implications

The application of quantum mechanics to resurrection raises profound philosophical and ethical questions.

1. The Nature of Identity

- If a person's quantum state is reconstructed, are they the same individual, or a new entity? Does continuity of identity require an unbroken stream of consciousness, or can it be reassembled from quantum information?

2. The Limits of Human Knowledge

- Quantum mechanics reveals the profound complexity and mystery of reality. Resurrection efforts must acknowledge the limits of human understanding and the potential for unintended consequences.

3. The Ethics of Manipulating Time

- Reversing or reconstructing the past challenges our conventional notions of morality, justice, and free will. What are the ethical implications of altering the past, and how should humanity navigate these responsibilities?

Conclusion

Quantum mechanics, with its mysterious and counterintuitive insights into time and reality, offers a tantalizing framework for understanding the possibility of universal scientific resurrection. By exploring phenomena such as superposition, entanglement, and time reversal, we glimpse the potential to bridge the gap between the past and the present, retrieving the essence of those who have passed.

However, the application of quantum principles to resurrection remains speculative, requiring not only scientific breakthroughs but also philosophical and ethical reflection. As humanity ventures into the unknown, quantum mechanics reminds us of the profound interconnectedness of all things and the boundless possibilities that await in the mysterious fabric of time. Through this lens, resurrection becomes not merely a technical challenge but a journey into the deepest questions of existence and our place in the cosmos.

Subsection 2: Biology, Death, and the Limits of Present Science

The human drive to overcome death has persisted throughout history, inspiring myths, religions, and now, scientific endeavors. Despite remarkable advances in medicine and biology, death remains one of the greatest mysteries of existence. Understanding it fully is essential to the project of universal scientific resurrection. To repair the past by resurrecting the dead, we must first grapple with the biological processes underlying life and death, and confront the limits of present science in reversing these processes.

This subsection examines the biological mechanisms of death, the challenges posed by the physical and informational decay of the human organism, and the current boundaries of scientific knowledge. It also explores the potential pathways by which future breakthroughs could overcome these limitations, bringing us closer to the dream of universal resurrection.

The Biology of Life and Death

Life is a highly organized, self-sustaining process. Biological systems maintain order and function through intricate networks of chemical reactions, cellular processes, and genetic information.

The cessation of these processes—death—marks the endpoint of life as we currently understand it.

1. The Mechanisms of Life

- At its core, life depends on the ability of cells to sustain themselves through metabolism, the conversion of energy into the molecules necessary for growth, repair, and reproduction.
- Genetic information, encoded in DNA, provides the blueprint for the development and maintenance of an organism. This information is faithfully replicated and passed on during cell division.
- Homeostasis, the ability to maintain stability in response to environmental changes, is a hallmark of life. When homeostasis breaks down irreversibly, the organism can no longer sustain life.

2. What Happens When We Die?

Death occurs when the biological processes that sustain life cease. The causes of death vary—disease, trauma, aging—but the endpoint is the same: the irreversible breakdown of cellular and systemic functions.

- Cellular Death: Cells die through processes like necrosis (uncontrolled cell death due to injury) or apoptosis (programmed cell death). In death, these processes cascade, leading to the collapse of tissues and organs.
- Brain Death: The irreversible cessation of brain activity is a clinical marker of death. Without brain function, consciousness and the ability to sustain bodily functions are lost.
- Decay and Decomposition: After death, the body undergoes decomposition, as cells and tissues break down due to enzymatic activity and microbial action. This leads to the loss of the physical and informational integrity of the organism.

3. Aging and Death

- Aging is a major contributor to death and is driven by the gradual accumulation of cellular damage over time. Key factors include:
- Genetic Mutations: Errors in DNA replication can accumulate, impairing cellular function.

- <u>Telomere Shortening</u>: Telomeres, protective caps on the ends of chromosomes, shorten with each cell division, eventually leading to cellular senescence.
- <u>Oxidative Stress</u>: Reactive oxygen species (ROS) generated during metabolism can damage DNA, proteins, and lipids.
- While aging is not the sole cause of death, it creates vulnerabilities that increase the likelihood of disease and system failure.

The Limits of Present Science

Despite significant advances in understanding biology and prolonging life, modern science remains far from reversing death. The challenges are immense, spanning both physical and informational dimensions.

1. Irreversible Biological Breakdown
- Once cellular processes cease, the body begins to degrade rapidly. The <u>biochemical and structural integrity</u> of cells, tissues, and organs deteriorates, making restoration increasingly difficult over time.
- Advanced stages of decomposition involve the complete breakdown of DNA and other biomolecules, erasing the blueprint of the individual's physical and genetic identity.

2. The Complexity of the Human Brain
- The brain is the seat of consciousness, memory, and identity. Its extreme complexity poses a significant challenge to resurrection.
- Neurons and their connections, known as <u>synapses</u>, encode memories and personality traits. When the brain is damaged or decayed, this intricate network is lost. Current neuroscience lacks the tools to fully map and restore these connections.

3. The Problem of Information Loss
- In addition to physical decay, death results in the loss of <u>informational integrity</u>—the unique data encoded in an individual's genetic material, neural patterns, and life experiences.

- Even if the body could be reconstructed, the absence of accurate information about the person's memories, personality, and consciousness would result in a mere physical replica, not a true resurrection.

4. Ethical and Technical Challenges in Cryonics
   - Cryonics, the preservation of the body at extremely low temperatures, is one of the few existing approaches aimed at preventing decay after death. However, it remains an experimental and controversial field.
   - Challenges include:
   - Cellular Damage: Ice crystal formation during freezing can destroy cells and tissues.
   - Incomplete Preservation: Current cryonic methods cannot preserve the full neural structure required for memory and identity restoration.
   - Lack of Reanimation Techniques: No existing technology can revive a cryonically preserved individual.

5. Current Limitations in Regeneration and Cloning
   - Regeneration of tissues and organs through stem cell therapy and 3D bioprinting has made significant progress, but these methods are far from being able to reconstruct an entire human body or brain.
   - Cloning can create genetic copies of an organism, but it does not replicate the unique memories, experiences, and personality of the original individual.

Potential Pathways Beyond the Limits

While present science cannot overcome death, several emerging fields and speculative technologies offer glimpses of how the limits might be transcended in the future.

1. Advances in Regenerative Medicine
   - Continued progress in stem cell research and tissue engineering could one day enable the regeneration of entire organs, including the brain.

- Gene editing technologies like CRISPR may allow scientists to repair genetic damage and reverse aging processes, prolonging life and potentially repairing the effects of death.

2. Digital Preservation of Identity
   - Efforts to digitally preserve human identity, such as mind uploading and neural mapping, aim to create a digital replica of a person's consciousness. While speculative, these technologies could serve as a bridge to future resurrection methods.

3. Quantum Information Retrieval
   - As explored in the previous subsection, quantum theories suggest that information may never be truly lost. Future technologies might retrieve the "quantum imprint" of a person's identity, allowing for their reconstruction.

4. Cryonics and Biostasis
   - Improvements in cryonics, including the development of vitrification techniques (which prevent ice formation), may enhance the preservation of the body and brain after death.
   - Biostasis technologies, such as suspended animation, could one day allow for long-term preservation without decay, buying time until resurrection methods are developed.

5. Artificial Intelligence and Reconstruction of the Past
   - Advanced artificial intelligence could analyze historical records, genetic material, and digital traces to reconstruct a person's identity, filling in gaps left by biological decay.
   - While controversial, this approach could complement physical resurrection methods by restoring lost memories and personality traits.

Philosophical and Ethical Considerations

The pursuit of resurrection raises profound philosophical and ethical questions, particularly in light of the limits of present science.

1. What Defines a Person?
- Is a person defined by their physical body, their memories, their consciousness, or some combination of these? If one aspect is restored but others are lost, is the resurrected individual truly the same person?

2. The Ethics of Experimentation
- Pushing the boundaries of science to overcome death may involve controversial experiments with human remains, genetic material, or consciousness. How should these experiments be regulated to ensure ethical integrity?

3. The Responsibility of Resurrection
- If future science enables resurrection, who decides who is resurrected, and under what conditions? What responsibilities do the living have to the resurrected, and how will society integrate them?

Conclusion

Biology provides the foundation for understanding life and death, but it also highlights the immense challenges of reversing death with current scientific knowledge. The limits of present science—irreversible decay, the complexity of the brain, and the loss of information—are significant barriers to resurrection. However, emerging fields such as regenerative medicine, quantum information theory, and artificial intelligence hold the promise of transcending these limitations in the future.

The journey to overcome death is not merely a scientific endeavor but a profound exploration of what it means to be human. By confronting the biological and philosophical challenges of resurrection, humanity takes its first steps toward repairing the past and honoring the lives of those who came before us. While the limits of present science remind us of the enormity of this task, they also inspire us to push beyond them, guided by the hope of a future where death is no longer the final word.

Section 2: Speculating on Breakthroughs

Subsection 1: Future Discoveries in Quantum Computing

Quantum computing represents one of the most transformative frontiers in modern science and technology. As humanity continues to push the boundaries of computation, the unique properties of quantum systems—superposition, entanglement, and quantum coherence—offer unprecedented potential to solve problems that are fundamentally intractable for classical computers. In the context of universal scientific resurrection, advanced quantum computing could play a pivotal role in overcoming challenges related to the retrieval, reconstruction, and simulation of the past.

This subsection explores the potential future discoveries in quantum computing that could enable the realization of resurrection. By examining the theoretical advancements, practical breakthroughs, and applications of quantum technologies, we gain insight into how quantum computing might bridge the gap between life and death, the past and the future.

The Unique Power of Quantum Computing

Quantum computers differ fundamentally from classical computers, which process information in binary bits (0s and 1s). Instead, quantum computers use qubits, which can exist in multiple states simultaneously due to the principle of superposition. This, combined with quantum entanglement and parallelism, allows quantum computers to process an immense number of possibilities at once.

1. Exponential Speedup for Complex Problems
   - Quantum computers excel at solving problems with exponential complexity, such as factoring large numbers (important for cryptography) or simulating molecular interactions.
   - In resurrection efforts, this computational power could enable the modeling and reconstruction of complex biological systems, including the human brain.

2. Quantum Parallelism

- Unlike classical computers, which solve problems step by step, quantum computers can evaluate multiple solutions simultaneously. This would be invaluable for sifting through vast datasets to reconstruct the identities and lives of deceased individuals.

3. The Role of Entanglement in Information Retrieval

- Quantum entanglement allows qubits to remain interconnected, even across large distances. Future quantum systems might leverage entanglement to retrieve and reconstruct information about the past, potentially reversing the effects of entropy and decay.

Applications of Quantum Computing to Resurrection

The unique capabilities of quantum computing could address some of the most daunting challenges in universal scientific resurrection. As the field advances, several specific applications stand out as transformative.

1. Reconstructing the Past Through Quantum Simulation

- Quantum computers could simulate the physical and informational states of deceased individuals by modeling the interactions of particles and systems at a fundamental level.
- Using data from preserved remains, historical records, or even traces left in the environment, quantum algorithms might recreate the molecular and neural structures of a person, effectively restoring their physical and cognitive essence.
- Reverse-Entropy Modeling: Quantum systems could use their computational power to simulate the reversal of entropy, reconstructing a person's state at a prior point in time.

2. Mapping and Restoring Neural Networks

- The human brain is an intricate network of approximately 86 billion neurons, each connected to thousands of others. Reconstructing this network is essential for restoring memory, personality, and consciousness.

- Quantum computing could enable the precise mapping and simulation of neural activity, identifying patterns that correspond to a person's unique mental and emotional characteristics.

- <u>Neural Entanglement Hypothesis</u>: If consciousness or memory has a quantum component (as some theories suggest), quantum computers might directly analyze and reconstruct these quantum states.

3. Decoding and Preserving Information

- Quantum computers could be used to retrieve and decode <u>lost information</u> about individuals, such as genetic data, neural patterns, or environmental imprints.

- Future quantum algorithms might recover fragmented or corrupted data, enabling the reconstruction of a person's identity even in the absence of complete biological remains.

4. Simulating Historical Contexts

- Resurrection is not merely about restoring individuals but also about placing them within their social, cultural, and historical contexts. Quantum computers could simulate entire historical environments with unparalleled accuracy.

- By modeling the interactions between resurrected individuals and their original surroundings, quantum systems could ensure that the resurrected are not just physical replicas but integrated into a meaningful narrative of their past lives.

Future Quantum Breakthroughs

Though quantum computing is still in its infancy, several anticipated breakthroughs could unlock its full potential for resurrection efforts.

1. Scalable and Fault-Tolerant Quantum Computers

- Current quantum computers are limited by the number of qubits and their susceptibility to errors caused by decoherence (loss of quantum states).

- Future advances in <u>quantum error correction</u> and scalable architectures could lead to robust quantum systems capable of handling the immense complexity of resurrection-related computations.

2. Quantum Artificial Intelligence (QAI)
- The integration of quantum computing with artificial intelligence could enable the development of quantum neural networks capable of learning and evolving.
- QAI systems could analyze vast amounts of data, identify patterns, and make decisions with a level of sophistication far beyond current AI technologies. This would accelerate the process of reconstructing identities and simulating consciousness.

3. Quantum Memory and Data Preservation
- Quantum memory systems, capable of storing and processing vast amounts of information in quantum states, could revolutionize data preservation.
- Future technologies might leverage quantum memory to store detailed information about individuals, ensuring that their identities can be reconstructed even after significant periods of time.

4. Entanglement Networks and Quantum Communication
- Advances in quantum networks could enable the development of global systems for sharing and processing quantum information.
- Such networks might facilitate the retrieval of information about deceased individuals from distributed sources, creating a unified framework for resurrection efforts.

Speculative Ideas: Quantum and the Nature of Time

Quantum computing's implications for resurrection extend beyond computation into the realm of time and causality, offering speculative yet intriguing possibilities.

1. Quantum Time Travel
- Some interpretations of quantum mechanics, such as the many-worlds interpretation or retrocausality, suggest that quantum systems can influence the past.

- Future quantum systems might exploit these phenomena to access and reconstruct past states of the universe, effectively "retrieving" the information needed for resurrection.

2. Quantum Holography
- Theoretical physics suggests that the universe may function as a hologram, with all information about its state encoded on a lower-dimensional surface (as in the holographic principle).
- Quantum computers could decode this holographic information, retrieving detailed records of every individual who has ever lived.

3. Universal Quantum Reconstruction
- If the universe is fundamentally a quantum system, advanced quantum computers might eventually simulate the entire cosmos, enabling the reconstruction of not only individual lives but entire historical epochs.
- This speculative idea aligns with the vision of resurrection as a universal project, repairing not just individual deaths but the fabric of history itself.

Challenges and Ethical Considerations

While the potential of quantum computing is vast, its application to resurrection raises significant challenges and ethical questions.

1. Technical Limitations
- Building and maintaining large-scale quantum computers is an immense technical challenge, requiring breakthroughs in materials science, engineering, and quantum mechanics.
- The complexity of reconstructing a human being—down to their memories, personality, and consciousness—may require levels of quantum computation far beyond what is currently conceivable.

2. The Authenticity of Resurrection
- If quantum computers reconstruct a person based on incomplete or approximate data, is the resulting individual truly the same as the original?

- The philosophical question of identity—whether a recreated person is a continuation of the original or a new entity—remains unresolved.

3. Ethical Use of Quantum Power
- The immense power of quantum computing could be misused, raising concerns about privacy, consent, and the potential for exploitation of resurrected individuals.
- Developing ethical guidelines for the use of quantum technologies in resurrection will be essential to ensure that this power is used responsibly.

Conclusion

Quantum computing represents a revolutionary leap in humanity's ability to process and manipulate information, offering transformative possibilities for universal scientific resurrection. From simulating the intricate structures of the human brain to reconstructing the past through quantum time reversal, the future discoveries in quantum computing could provide the tools needed to repair the past and restore the dead.

However, the journey from theory to application will require unprecedented advances in science, technology, and ethics. The challenges are immense, but so too is the promise of quantum computing to unlock the mysteries of life, death, and time itself. As humanity ventures into this new frontier, quantum computing may become the cornerstone of a future where the past is no longer irretrievably lost and where death is no longer the final boundary.

Subsection 2: Harnessing AI to Decode Historical and Biological Data

The dream of universal scientific resurrection hinges on the ability to gather, interpret, and reconstruct vast amounts of data about individuals who once lived. This data encompasses both biological information—such as DNA, cellular structures, and neural patterns—and historical information, including personal artifacts, written records, and the social and cultural contexts of their lives. The sheer scale and complexity of this undertaking far exceed the capabilities of traditional methods of analysis.

Enter artificial intelligence (AI), a transformative technology designed to process and analyze massive and complex datasets with unprecedented speed and precision. AI has already revolutionized fields such as medicine, archaeology, and historical research, and its future advancements hold the potential to bridge the gaps in our knowledge of the past. By harnessing AI, humanity can decode and reconstruct the information necessary for resurrection, combining fragmented biological and historical data into a coherent picture of the individuals we seek to restore.

The Role of AI in Universal Resurrection

AI's ability to analyze patterns, make predictions, and simulate complex systems makes it uniquely suited to the challenges of resurrection. Below are key areas where AI can contribute to decoding and reconstructing both biological and historical data.

1. Decoding Biological Data

Biological data forms the foundation for reconstructing the physical and cognitive essence of a person. AI has already demonstrated remarkable capabilities in understanding the intricacies of life at the molecular and cellular levels, and future developments will take this even further.

1. Reconstructing Genomic Information
  - AI-powered algorithms can analyze fragments of DNA, even from highly degraded samples, to reconstruct complete genomes.
  - Advanced AI models, trained on large datasets of genetic sequences, could infer missing segments of DNA, filling in gaps caused by decay or damage over time.
  - AI tools like deep learning can also identify and correct errors in sequencing data, ensuring accurate reconstructions of an individual's genetic blueprint.

- 182 -

2. Simulating Cellular and Organ Function
- AI can simulate the behavior of cells, tissues, and organs, enabling the virtual reconstruction of biological systems.
- For example, AI-driven models could predict how a person's cells would interact and regenerate, offering insights into how their body might be physically restored.
- By integrating biological data with AI simulations, researchers could even recreate specific physiological features, such as immune responses or metabolic processes, that are unique to an individual.

3. Mapping Neural Structures and Consciousness
- The restoration of a person's consciousness, memories, and personality requires a detailed understanding of the brain's neural networks. AI is already being used to map the brain at unprecedented levels of detail through projects like the Human Connectome Project.
- In the future, AI could analyze neural decay patterns and use this information to reconstruct the synaptic connections that encode an individual's thoughts, emotions, and memories.
- AI models might also simulate the dynamic processes of the brain, bridging the gap between physical structure and cognitive function to restore a person's unique sense of self.

4. Synthesizing Missing Data Through Prediction
- In cases where biological data is incomplete, AI could infer missing information by analyzing patterns in related datasets. For example:
- AI could use family genetic data to reconstruct gaps in an individual's genome.
- Neural network models might predict missing connections in damaged brain scans, helping to restore lost memories or personality traits.

2. Decoding Historical and Personal Data

In addition to biological reconstruction, resurrection requires placing individuals within their historical and cultural contexts. AI's ability to analyze and interpret vast amounts of historical data makes it a powerful tool for uncovering the stories and identities of those who have passed.

1. Analyzing Artifacts and Cultural Records
- AI can process and analyze historical artifacts, texts, photographs, and other records to reconstruct the lives of individuals.
- For example, natural language processing (NLP) algorithms like GPT-style models can analyze historical documents to extract details about a person's life, such as their occupation, relationships, and significant events.
- AI systems trained on cultural datasets could also recreate the social and historical environments in which individuals lived, providing context for their resurrection.

2. Reconstructing Personal Histories
- AI can use fragmented data—such as letters, diaries, or digital traces left behind by modern individuals—to piece together detailed personal histories.
- Machine learning models could analyze patterns in writing, speech, or behavior to infer personality traits, preferences, and emotional states.
- By combining multiple sources of data, AI could create a comprehensive narrative of a person's life, ensuring that their resurrection reflects not only their physical form but also their unique identity.

3. Simulating Historical Contexts
- Resurrection is not just about restoring individuals but also about integrating them into meaningful contexts. AI could simulate the social, cultural, and ecological environments in which individuals once lived.
- For example, AI-driven virtual reality (VR) systems could recreate historical cities, communities, or ecosystems, allowing resurrected individuals to experience life as it once was.
- These simulations could also help prepare resurrected individuals for modern life, bridging the gap between their original context and the present day.

4. Filling Gaps in Historical Knowledge
   - AI has already been used to decode ancient languages, reconstruct lost artworks, and identify historical patterns. Future AI systems could fill in missing details about individuals whose records are incomplete.
   - By analyzing large-scale historical datasets, AI could infer connections between individuals, events, and cultures, providing a richer understanding of the past.

3. Integrating Biological and Historical Data

One of the greatest challenges of resurrection is uniting biological and historical data into a cohesive whole. AI, with its ability to process and integrate complex datasets, will be essential in this effort.

1. Creating Holistic Profiles
   - AI could combine genetic data, neural maps, and historical records to create a holistic profile of an individual.
   - This profile would include not only their physical form but also their memories, personality, and cultural identity, ensuring that the resurrected individual is as complete as possible.

2. Simulating Lived Experiences
   - AI could simulate the lived experiences of individuals by integrating biological and historical data. For example:
   - Neural simulations might recreate how a person responded to specific events in their life.
   - Historical simulations could place individuals in scenarios that reflect their original environment, helping to refine and validate their reconstruction.

3. Resolving Ambiguities
   - In cases where data is contradictory or incomplete, AI could use probabilistic models to resolve ambiguities.
   - For instance, if two conflicting accounts of a person's life exist, AI could analyze the likelihood of each scenario based on broader historical or biological patterns.

Future AI Breakthroughs

To fully realize the potential of AI in resurrection efforts, several key advancements will be necessary:

1. Advanced Machine Learning Models
   - Future AI systems will need to process and analyze data with greater accuracy, speed, and contextual understanding.
   - New algorithms capable of unsupervised learning could identify patterns and relationships in data without human guidance, uncovering hidden insights about the past.

2. AI-Driven Neural Emulation
   - Emulating the neural activity that underpins consciousness and memory will require highly sophisticated AI models capable of simulating the brain's dynamic processes.
   - These models will need to integrate data from multiple sources, including biological scans and historical records, to recreate a person's cognitive and emotional life.

3. Ethical AI Frameworks
   - As AI takes on a central role in resurrection, it will be essential to develop ethical frameworks to guide its use.
   - These frameworks should address questions of consent, privacy, and the authenticity of reconstructed individuals, ensuring that resurrection is conducted responsibly.

Ethical and Philosophical Considerations

Harnessing AI for resurrection raises profound ethical and philosophical questions:

1. Accuracy and Authenticity
   - How accurate does a reconstruction need to be for it to count as a true resurrection?
   - If AI "fills in the gaps" with probabilistic predictions, does the result still reflect the original individual?

2. Consent and Agency
   - Can individuals who lived in the past meaningfully consent to being resurrected?
   - How should resurrected individuals be treated, and what rights should they have?

3. The Role of AI in Shaping History
   - By reconstructing the past, AI has the power to influence how history is remembered and understood.
   - How can we ensure that AI-driven reconstructions remain faithful to the truth, rather than reflecting the biases of their creators?

Conclusion

Artificial intelligence offers humanity the tools to decode the mysteries of both biology and history, making universal scientific resurrection a tangible possibility. By analyzing fragmented data, reconstructing neural and genetic structures, and simulating historical contexts, AI can bridge the gap between what was lost and what might be restored.

However, as we harness the power of AI, we must also grapple with the ethical and philosophical implications of resurrection. In doing so, we ensure that this transformative technology is used not only to repair the past but to honor the dignity and humanity of those we seek to bring back. Through AI, the boundaries between life, death, and time may one day blur, allowing us to fulfill the audacious dream of restoring the dead to life.

Part V: Toward the Realization of the Beloved Community

Chapter 9: Building the Future

Section 1: From Individual Resurrection to Universal Redemption

Subsection 1: The Role of Rational Hope in Human Progress

Throughout history, <u>hope</u> has been a driving force behind humanity's greatest achievements. It is the belief in the possibility of a better future that inspires individuals and societies to strive for progress, even in the face of overwhelming challenges. However, not all hope is created equal. While <u>blind hope</u> may lead to unrealistic expectations or false comfort, <u>rational hope</u>—grounded in evidence, reason, and a willingness to work toward ambitious goals—has the power to transform the world.

In the context of <u>universal scientific resurrection</u>, rational hope plays a critical role in envisioning a future where the boundaries of life and death can be transcended. This subsection explores the concept of rational hope, its importance in driving scientific and philosophical progress, and how it serves as a cornerstone for the audacious goal of repairing the past and resurrecting the dead.

<u>What is Rational Hope?</u>

Rational hope is the belief in the possibility of achieving meaningful outcomes, tempered by a realistic understanding of the obstacles involved. It combines <u>aspiration with critical thinking</u>, ensuring that hope is not mere wishful thinking but a deliberate and purposeful attitude.

1. <u>The Key Elements of Rational Hope</u>
  - <u>Vision</u>: Rational hope begins with a clear and inspiring vision of what could be possible, such as the eradication of disease, the expansion of human knowledge, or the resurrection of lost individuals.
  - <u>Evidence-Based Optimism</u>: Unlike blind hope, rational hope is grounded in evidence, relying on scientific discoveries,

technological advancements, and historical trends to justify its plausibility.

- Resilience and Determination: Rational hope acknowledges the difficulties and uncertainties involved but remains steadfast in the belief that progress, though challenging, is attainable.

2. Hope vs. Blind Optimism

- Blind optimism assumes that good outcomes will occur regardless of effort or evidence, often ignoring the complexities of reality.
- Rational hope, by contrast, embraces the complexity of the world and requires active engagement with the challenges at hand. It is both aspirational and pragmatic, rooted in the understanding that meaningful progress requires hard work, collaboration, and ingenuity.

The Historical Power of Rational Hope

Rational hope has been a guiding principle in many of humanity's greatest endeavors, driving progress in science, technology, and social development. It has allowed people to envision and work toward futures that seemed impossible in their own time.

1. Scientific Breakthroughs

- The history of science is filled with examples of individuals and communities pursuing bold ideas despite skepticism or uncertainty:
- Space Exploration: The moon landing in 1969 exemplifies rational hope. While the obstacles were immense, the vision of exploring space was grounded in scientific innovation and meticulous planning.
- Medical Advancements: The eradication of smallpox, the development of antibiotics, and the decoding of the human genome were all fueled by the belief that science could overcome the challenges posed by disease and biology.

2. Social and Ethical Progress
- Rational hope has also driven movements for social justice and human rights, inspiring efforts to create a more equitable and compassionate world:
  - The abolition of slavery, the fight for gender equality, and the civil rights movement were rooted in the belief that humanity could transcend its historical injustices and create a better future.

3. Overcoming Existential Challenges
- Humanity has faced numerous existential threats, from pandemics to world wars, yet rational hope has enabled societies to persevere. By focusing on evidence-based solutions and collective action, people have repeatedly demonstrated their capacity to adapt and thrive.

Rational Hope and the Vision of Universal Scientific Resurrection

The idea of resurrecting the dead may seem audacious, even fantastical, to many. However, when viewed through the lens of rational hope, it becomes an inspiring and plausible goal rooted in humanity's ability to innovate and solve complex problems.

1. The Basis for Hope in Resurrection
- Scientific and technological advancements provide a foundation for rational hope in universal resurrection:
  - Biotechnology: The rapid progress in genetic engineering, tissue regeneration, and neuroscience suggests that the restoration of life may one day be within reach.
  - Quantum Computing and AI: Emerging technologies capable of handling vast amounts of data and simulating complex systems offer tools for reconstructing the identities and physical forms of deceased individuals.
  - Information Preservation: Theories in physics, such as the conservation of information, imply that the essence of a person's existence—encoded in their biology and environment—may not be lost forever.

2. Hope as a Catalyst for Action
- Rational hope in resurrection inspires action by encouraging people to invest in research, collaborate across disciplines, and overcome the ethical and technical challenges involved.
- It also fosters a sense of purpose and meaning, as the effort to repair the past represents a profound act of justice and reconciliation.

3. Balancing Hope with Realism
- While rational hope embraces the possibility of resurrection, it also acknowledges the uncertainties and limitations of current science. Achieving such a goal will require generations of effort, creativity, and perseverance.

The Ethical Dimension of Rational Hope

Rational hope is not merely a tool for achieving progress; it is also a moral stance that reflects humanity's capacity for compassion, responsibility, and a commitment to the greater good.

1. Honoring the Past
- The pursuit of resurrection represents an acknowledgment of the value and dignity of those who have come before us. Rational hope drives the belief that their lives were not meaningless and that their restoration is a noble endeavor.

2. A Responsibility to Future Generations
- Rational hope extends beyond the present, recognizing our responsibility to create a better world for future generations. By investing in the possibility of resurrection, humanity demonstrates its commitment to overcoming the limitations of mortality and expanding the horizons of human potential.

3. Avoiding Hubris
- While rational hope encourages ambitious goals, it must be tempered by humility and ethical reflection. The pursuit of resurrection must be guided by respect for the complexity of life and the limits of human knowledge.

Fostering Rational Hope in the Modern Age

In an era marked by rapid technological change and existential challenges, fostering rational hope is more important than ever. To sustain the effort required for universal resurrection, individuals and societies must cultivate a hopeful yet realistic mindset.

1. Education and Public Engagement
- Promoting scientific literacy and critical thinking helps people understand the evidence and reasoning behind ambitious goals like resurrection.
- Public engagement fosters a sense of collective purpose, encouraging collaboration and investment in long-term projects.

2. Celebrating Progress and Possibility
- Highlighting achievements in science and technology reinforces the belief that humanity is capable of overcoming even the most daunting challenges.
- Sharing stories of progress inspires future generations to dream big and pursue bold ideas.

3. Building a Culture of Resilience
- Rational hope thrives in a culture that values resilience and perseverance. By embracing setbacks as opportunities for learning and growth, societies can maintain their commitment to ambitious goals.

Conclusion

Rational hope is the foundation of human progress, combining the power of imagination with the rigor of evidence-based reasoning. It has driven humanity to explore the unknown, solve complex problems, and create a better future for all. In the context of universal scientific resurrection, rational hope serves as both an inspiration and a guide, encouraging humanity to strive for the seemingly impossible while remaining grounded in the realities of science and ethics.

By fostering rational hope, we honor the past, embrace the present, and build a future in which the boundaries of life and death are no longer insurmountable. It is this hope that propels us forward, reminding us that the greatest achievements begin with the audacious belief that progress is possible—even in the face of mortality itself. Through rational hope, humanity can dare to ask, "Shall the future repair the past?" and take the first steps toward making the answer a resounding "yes."

Subsection 2: Transforming Impossible Dreams into Reality

History is rich with examples of ideas once deemed impossible that, through perseverance, ingenuity, and the relentless pursuit of knowledge, became reality. The dream of universal scientific resurrection—the restoration of those who have passed—stands at the crossroads of imagination and emerging science. Transforming such an audacious vision into reality will require not only breakthroughs in technology but also a paradigm shift in how humanity approaches problems of extraordinary complexity.

This subsection explores how humanity has historically turned impossible dreams into achievements, identifies the key ingredients for overcoming the barriers to resurrection, and outlines a roadmap for transforming this bold vision into a tangible future.

The Historical Precedent: From Impossibility to Achievement

The progress of civilization has often been fueled by ideas that initially seemed unattainable. These once-impossible dreams serve as reminders that impossibility is often a reflection of the present, not the future.

1. The Impossible Becomes Inevitable
   - Flight: For millennia, humans dreamed of taking to the skies. While it seemed impossible in ancient times, the Wright brothers' first powered flight in 1903 was the culmination of centuries of scientific progress. Today, air travel is a routine part of life.

- Space Exploration: The idea of humans walking on the moon was dismissed as science fiction for much of history. Yet, through the Apollo program, humanity achieved this feat in 1969, driven by an unyielding commitment to push beyond perceived boundaries.
- Curing Diseases: Diseases that were once death sentences—such as smallpox and polio—have been eradicated or controlled through the development of vaccines and medical science, achievements that seemed impossible just a few centuries ago.

2. The Common Thread
- What unites these accomplishments is a combination of visionary leadership, scientific rigor, technological innovation, and the determination to overcome obstacles.
- These examples illustrate that what seems impossible today may simply require the right convergence of knowledge, tools, and collective will.

The Challenges of Universal Resurrection

Resurrecting the dead is one of humanity's most ambitious dreams, and the challenges involved are extraordinary. However, by understanding these obstacles, we can begin to chart a course toward overcoming them.

1. Scientific and Technical Barriers
- Reconstructing Decayed Information: The physical decay of bodies after death leads to the loss of biological and informational integrity. Reconstructing DNA, neural networks, and personality traits demands technologies capable of retrieving and synthesizing missing data.
- Complexity of Consciousness: The human mind is an intricate web of memories, emotions, and experiences. Restoring consciousness requires a deep understanding of how these elements arise from the brain's structure and function.
- Integration of Diverse Fields: Resurrection efforts require breakthroughs across multiple disciplines, including quantum computing, artificial intelligence, regenerative medicine, and theoretical physics.

2. Philosophical and Ethical Concerns
- The Nature of Identity: If a person's body, mind, and memories are reconstructed, are they truly the same individual, or a new entity entirely?
- Consent and Autonomy: Should humanity resurrect individuals without their consent, especially those who lived in times when such technologies were unimaginable?
- Social and Cultural Impact: How would resurrected individuals adapt to modern society, and how would their return affect current generations?

3. Cultural and Psychological Barriers
- Skepticism and Resistance: Many may dismiss resurrection as an unrealistic fantasy, hindering investment and research.
- Fear of Playing God: The idea of reversing death challenges deeply held religious, ethical, and philosophical beliefs about the natural order of life.

The Path Forward: Ingredients for Transforming Dreams into Reality

To turn the impossible into reality, humanity must cultivate the conditions necessary for innovation and perseverance. The following elements are critical for transforming universal scientific resurrection from a dream into a practical achievement.

1. A Clear and Compelling Vision
- Great achievements begin with a bold and inspiring vision. The goal of resurrection must be framed not as a fantastical pursuit but as a profound act of healing, justice, and honoring the past.
- Visionary thinkers and leaders must articulate and champion this goal, galvanizing resources and minds to work toward it.

2. Interdisciplinary Collaboration
- The complexity of resurrection demands collaboration across fields that rarely intersect. For example:
- Biologists and neuroscientists must work with physicists to understand the nature of information and consciousness.
- AI experts must collaborate with historians to reconstruct personal identities and cultural contexts.

- Fostering a culture of interdisciplinary research will enable breakthroughs that no single field could achieve alone.

3. Incremental Progress and Milestones
- Transforming impossible dreams into reality requires breaking them down into achievable steps. For example:
- Short-Term Goals: Advances in cryonics, tissue regeneration, and neural mapping.
- Medium-Term Goals: Developing AI systems capable of reconstructing incomplete data and simulating consciousness.
- Long-Term Goals: Integrating these technologies to achieve full-fledged resurrection.
- Celebrating milestones along the way reinforces the belief that progress is possible.

4. Funding and Institutional Support
- Significant achievements require sustained investment in research and development. Governments, private institutions, and philanthropists must recognize the value of resurrection research and commit resources to its pursuit.
- Establishing dedicated research centers and fostering public-private partnerships can accelerate progress.

5. A Culture of Resilience and Optimism
- Transforming impossible dreams into reality often involves setbacks and failures. A culture that values resilience and optimism can help researchers and innovators persevere in the face of challenges.
- By framing resurrection as a moral and scientific imperative, society can build the collective will necessary to sustain long-term efforts.

6. Ethical Frameworks and Dialogue
- As progress is made, ethical considerations must remain central to the pursuit of resurrection. Open dialogue among scientists, ethicists, religious leaders, and the public can help navigate the moral complexities involved.

- Developing clear ethical guidelines ensures that resurrection is pursued responsibly and with respect for the dignity of those it seeks to restore.

Inspiration from Today's Transformative Technologies

Modern technological advancements offer a glimpse of how seemingly impossible dreams can become reality. The following examples highlight the transformative potential of human ingenuity:

1. Artificial Intelligence
   - AI systems like GPT-4 have demonstrated the ability to process and generate human-like text, analyze patterns in vast datasets, and simulate complex systems. Future iterations of AI could play a central role in reconstructing individuals from fragmented data.

2. CRISPR and Genetic Engineering
   - Tools like CRISPR have revolutionized the ability to edit DNA, enabling scientists to correct genetic disorders and potentially reverse aging processes. These technologies lay the groundwork for restoring the biological aspects of deceased individuals.

3. Quantum Computing
   - Emerging quantum computers have the potential to simulate biological processes, decode vast amounts of information, and even model the physics of time and entropy. These capabilities are critical for addressing the informational challenges of resurrection.

4. Virtual Reality and Simulation
   - Advances in virtual reality and simulation technologies are creating immersive environments that could one day reconstruct the historical and cultural contexts of resurrected individuals, helping them reintegrate into society.

Conclusion

Transforming impossible dreams into reality is one of humanity's defining traits. The vision of universal scientific resurrection, while audacious, is grounded in the same spirit of curiosity, determination, and innovation that has driven progress throughout history. By embracing a clear vision, fostering interdisciplinary collaboration, and cultivating resilience in the face of challenges, humanity can work toward making this dream a reality.

As history has shown, the line between the impossible and the achievable is often drawn by the limits of imagination and effort. Universal resurrection challenges us to expand those limits, to repair the past, and to honor the lives of those who came before us. It is a profound and noble pursuit—one that reminds us that the greatest achievements begin with the courage to dream and the resolve to act. Through this lens, resurrection is not just a scientific endeavor but a testament to humanity's enduring capacity to transform the impossible into the inevitable.

Section 2: The Long-Term Vision

Subsection 1: Creating Sustainable Frameworks for Resurrection

The pursuit of universal scientific resurrection—the restoration of individuals who have passed—requires more than technological breakthroughs and scientific ingenuity. It demands the establishment of sustainable frameworks that balance the practical, ethical, and social dimensions of this audacious goal. Without a robust foundation to support resurrection efforts, the endeavor risks becoming unsustainable, ethically fraught, or inaccessible to future generations.

This subsection explores the key components of creating sustainable frameworks for resurrection, focusing on the integration of technology, societal systems, ethical considerations, and environmental responsibility. By addressing these factors, humanity can ensure that the pursuit of resurrection is not only achievable but also equitable, responsible, and enduring.

The Importance of Sustainability in Resurrection

Resurrection is an inherently ambitious and resource-intensive pursuit, involving complex technologies, long timelines, and profound ethical implications. Sustainability is essential to ensure that these efforts can be carried out in a way that benefits society as a whole without depleting resources or causing harm.

1. Technological Longevity
- Resurrection will require advanced technologies in fields such as biotechnology, artificial intelligence, and quantum computing. These systems must be designed to operate reliably over long periods, as the process of developing and implementing resurrection techniques may span decades or even centuries.

2. Ethical Responsibility
- Creating frameworks for resurrection involves addressing profound questions about identity, consent, and the moral implications of restoring life. Sustainable frameworks must incorporate ethical principles to guide decision-making and ensure that resurrection efforts respect the dignity and autonomy of individuals.

3. Social Equity
- Resurrection must not become an exclusive privilege available only to the wealthy or powerful. Sustainable frameworks should prioritize accessibility and fairness, ensuring that the benefits of resurrection extend to all of humanity, regardless of socioeconomic status.

4. Environmental Impact
- The resources required for resurrection—energy, materials, and infrastructure—must be managed responsibly to minimize environmental impact. Sustainable practices are essential to ensure that resurrection efforts do not exacerbate existing ecological challenges.

Key Components of Sustainable Frameworks for Resurrection

To build a sustainable foundation for resurrection, several key components must be addressed. These include technological infrastructure, governance systems, ethical guidelines, and environmental considerations.

1. Building Scalable and Resilient Technological Systems

The technologies required for resurrection must be scalable, reliable, and adaptable to future advancements.

1. Interdisciplinary Research Hubs
- Establishing global research hubs dedicated to resurrection-related sciences (e.g., biomedicine, AI, quantum computing) can foster collaboration across disciplines.
- These hubs should share knowledge and resources, ensuring that progress is distributed equitably and not concentrated in a few regions or institutions.

2. Modular and Scalable Infrastructure
- Resurrection technologies should be designed with modularity in mind, allowing them to scale as resources and knowledge expand.
- For example, AI systems for reconstructing neural networks or DNA could start by focusing on small-scale projects and gradually expand to handle larger datasets and more complex tasks.

3. Long-Term Data Preservation
- Preserving biological and historical data is critical for successful resurrection. Sustainable frameworks must include robust systems for storing and protecting data over long periods.
- Technologies such as DNA data storage and quantum memory could provide reliable, long-term solutions for preserving the information needed for resurrection.

4. Energy-Efficient Technologies
- The energy demands of resurrection technologies, particularly quantum computing and AI, could be immense. Developing

energy-efficient systems is essential to ensure that resurrection efforts are sustainable and do not contribute to environmental degradation.

2. Establishing Governance and Global Collaboration

Resurrection is a global endeavor that requires cooperative governance and shared responsibility. Sustainable frameworks must include systems for decision-making, regulation, and accountability.

1. International Cooperation
   - Resurrection should be treated as a collective human project, with nations and organizations working together to pool resources and expertise.
   - An international body, similar to the United Nations, could oversee resurrection efforts, ensuring that they align with global priorities and ethical standards.

2. Transparent Decision-Making
   - Governance systems must be transparent and inclusive, incorporating input from scientists, ethicists, policymakers, and the public.
   - Decisions about who is resurrected, how resources are allocated, and what technologies are developed should be made through open, democratic processes.

3. Equitable Access
   - Sustainable frameworks must ensure that resurrection is accessible to all, regardless of geography, socioeconomic status, or historical context.
   - Policies should be developed to prioritize equity, such as subsidizing resurrection efforts for marginalized communities or individuals from underrepresented populations.

4. Legal and Ethical Guidelines
   - Governance systems must establish clear legal and ethical guidelines to address questions such as:
     - What criteria determine eligibility for resurrection?

- How should resurrected individuals be reintegrated into society?
- What rights and responsibilities do resurrected individuals have?

3. Ethical Considerations and Public Engagement

Ethics must be at the heart of sustainable resurrection frameworks to ensure that the pursuit of this goal respects human dignity and aligns with societal values.

1. Consent and Autonomy
- A key ethical challenge is determining how to address the issue of consent for individuals who lived in times when resurrection was inconceivable.
- Sustainable frameworks should prioritize autonomy, allowing resurrected individuals to make decisions about their lives and futures.

2. Balancing Individual and Collective Needs
- Resurrection may require significant societal resources. Ethical frameworks must balance the needs of resurrected individuals with the well-being of the broader community.

3. Public Dialogue and Education
- Engaging the public in discussions about resurrection ensures that societal values and concerns are incorporated into decision-making.
- Education initiatives can help demystify resurrection technologies and build public trust and support for sustainable frameworks.

4. Minimizing Environmental Impact

Resurrection efforts must align with global priorities for sustainability, including the need to address climate change and resource conservation.

1. Energy Sustainability
- Developing renewable energy sources to power resurrection technologies—such as solar, wind, and fusion energy—will reduce their environmental footprint.
- Energy-efficient practices, such as optimizing computing processes and minimizing waste, are essential for sustainability.

2. Resource Conservation
- Resurrection frameworks should prioritize the use of sustainable materials and minimize the consumption of finite resources.
- Recycling and repurposing technologies can help reduce the environmental impact of resurrection infrastructure.

3. Integration with Ecological Restoration
- Resurrection efforts could contribute to ecological restoration by incorporating technologies that benefit the environment. For example:
  - AI-driven systems could be used to monitor and protect ecosystems.
  - Biotechnologies developed for resurrection could also be applied to conservation and biodiversity efforts.

5. Preparing Society for Resurrection

Sustainable frameworks must address the social and cultural implications of resurrection, ensuring that society is prepared to embrace this transformative goal.

1. Reintegration Programs
- Resurrected individuals will need support to adapt to modern society, including education, cultural orientation, and psychological counseling.
- Sustainable frameworks should include programs to help resurrected individuals integrate into communities and contribute meaningfully to society.

2. Cultural and Historical Preservation
- Resurrection frameworks should prioritize the preservation of cultural and historical knowledge, ensuring that resurrected individuals can reconnect with their heritage.

3. Fostering a Culture of Inclusivity
- Society must cultivate a culture of inclusivity and compassion, embracing resurrected individuals as valuable members of the human family.

Conclusion

Creating sustainable frameworks for resurrection is not merely a technical challenge; it is a profound moral and societal undertaking. By addressing the technological, ethical, social, and environmental dimensions of this goal, humanity can ensure that resurrection efforts are responsible, equitable, and enduring.

Sustainability is the foundation upon which the dream of universal resurrection can become a reality. It reflects humanity's commitment to honoring the past, embracing the present, and building a future in which the barriers of mortality are transcended—not recklessly, but thoughtfully and ethically. Through sustainable frameworks, we can transform the dream of repairing the past into a legacy that endures for generations to come.

Subsection 2: Envisioning the Beloved Community as a Cosmic Goal

The dream of universal scientific resurrection extends far beyond the restoration of individual lives. It represents a profound opportunity to reshape the human condition and create a world rooted in justice, compassion, and the recognition of our shared humanity. This vision aligns with the concept of the Beloved Community, first popularized by philosopher-theologian Josiah Royce and later adopted and expanded by Dr. Martin Luther King Jr.

The Beloved Community envisions a society in which the barriers of inequality, division, and suffering are dismantled, giving way to a universal fellowship defined by love, inclusion, and mutual respect. In the context of resurrection, this ideal takes on an even greater scope, transcending Earth and anchoring itself as a cosmic goal—a vision for humanity's role in the universe. This subsection explores how resurrection could contribute to the realization of this ideal and how the Beloved Community might evolve into a guiding principle for a compassionate and interconnected future.

The Concept of the Beloved Community

The Beloved Community is an ethical and spiritual ideal that calls for the transformation of human relationships and institutions. It is characterized by justice, reconciliation, and collective well-being, where the worth and dignity of every individual are recognized and upheld.

1. Core Principles of the Beloved Community
   - Radical Inclusivity: The Beloved Community embraces all individuals, regardless of race, class, gender, religion, or nationality, fostering a sense of unity and shared purpose.
   - Nonviolence and Reconciliation: Conflicts are resolved through understanding and compassion rather than coercion or violence.
   - Shared Prosperity: Economic and social systems are organized to ensure that no one is left behind, promoting equity and mutual care.

2. From Earthly Vision to Cosmic Aspiration
   - While traditionally framed as a societal goal on Earth, the Beloved Community can be expanded to encompass humanity's place in the cosmos.
   - The pursuit of resurrection, with its emphasis on repairing the past and honoring the lives of those who came before us, aligns with the ethical and spiritual values of the Beloved Community, extending its reach to all of existence.

Resurrection as a Pathway to the Beloved Community

The process of resurrecting individuals is not merely a technical or scientific endeavor—it is a deeply ethical and communal act, one that reflects humanity's capacity for compassion and justice. Universal resurrection could serve as a cornerstone for building the Beloved Community on a cosmic scale.

1. Healing Historical Injustices

The Beloved Community seeks to address and repair the wounds of the past. Resurrection offers a unique opportunity to confront the legacies of inequality, violence, and oppression that have shaped human history.

1. Restoring Lives Lost to Injustice
- Many lives throughout history were cut short by war, slavery, genocide, and systemic inequality. Resurrection provides a means to restore these individuals, honoring their inherent dignity and allowing them to reclaim the futures that were denied to them.
- This act of restoration is a profound gesture of reconciliation, acknowledging the suffering of the past while affirming humanity's commitment to justice and healing.

2. Creating a More Inclusive Narrative of History
- By resurrecting individuals from diverse backgrounds and time periods, humanity can create a richer and more inclusive narrative of history, one that recognizes the contributions and experiences of all people.
- This expanded understanding of history can foster greater empathy and appreciation for the interconnectedness of human lives.

2. Fostering Universal Compassion

The act of resurrection embodies the values of compassion and interconnectedness, reminding humanity of its shared responsibility to care for one another.

1. Recognizing the Dignity of Every Individual
   - Resurrection affirms that every life matters, regardless of its circumstances or historical context. By committing to the restoration of all individuals, humanity demonstrates its belief in the intrinsic worth of every person.
   - This recognition of universal dignity is a foundational principle of the Beloved Community, fostering a culture of mutual respect and care.

2. Expanding Empathy Across Time and Space
   - Resurrection challenges humanity to expand its circle of empathy, not only to those who are currently alive but also to those who have lived in the past and may live in the future.
   - This expanded perspective encourages a deeper sense of solidarity and interconnectedness, reflecting the values of the Beloved Community.

3. Building a Collective Future

The Beloved Community is not only about addressing the past but also about creating a future in which all individuals can thrive. Resurrection contributes to this vision by fostering collaboration, innovation, and shared purpose.

1. Reimagining Society
   - Resurrection invites humanity to rethink its social, economic, and political structures, ensuring that they are inclusive and equitable for all, including resurrected individuals.
   - This reimagining of society aligns with the principles of the Beloved Community, creating systems that prioritize collective well-being over individual gain.

2. Cultivating a Sense of Shared Destiny
   - The pursuit of resurrection represents a collective endeavor that transcends individual interests, uniting humanity around a shared goal.
   - This sense of shared destiny can inspire greater cooperation and a commitment to the common good, reinforcing the values of the Beloved Community.

The Cosmic Beloved Community

As humanity expands its horizons beyond Earth, the principles of the Beloved Community take on a cosmic significance. Resurrection, with its emphasis on repairing the past and honoring the interconnectedness of life, offers a framework for envisioning humanity's role in the universe.

1. A Universal Ethic of Care
- The Beloved Community calls for the extension of care and compassion to all beings. In the context of a cosmic vision, this ethic could be extended to include not only human beings but also other forms of life and the ecosystems that sustain them.
- Resurrection exemplifies this ethic, reflecting humanity's commitment to preserving and restoring the richness of life.

2. Bridging the Temporal Divide
- Resurrection bridges the divide between past, present, and future, creating a sense of continuity and shared existence across time.
- This temporal interconnectedness reflects the cosmic scope of the Beloved Community, emphasizing the unity of all life across space and time.

3. Humanity's Role in the Universe
- By pursuing resurrection and fostering the values of the Beloved Community, humanity can define its role in the cosmos as a force for restoration, reconciliation, and flourishing.
- This vision challenges humanity to rise above narrow self-interest and embrace its potential as a steward of life and a builder of harmony on a universal scale.

Challenges and Opportunities

While the vision of the Beloved Community as a cosmic goal is inspiring, it also poses significant challenges. Addressing these challenges is essential to ensuring that resurrection efforts align with the values of justice, compassion, and inclusivity.

1. Preventing Inequity and Exploitation
- Resurrection must not become a tool for reinforcing existing inequalities or consolidating power in the hands of a few. Sustainable frameworks and ethical guidelines are essential to ensure that its benefits are distributed equitably.

2. Navigating Cultural and Philosophical Diversity
- Humanity is diverse, with differing beliefs, values, and traditions. Building the Beloved Community requires finding common ground while respecting this diversity.

3. Sustaining Hope and Commitment
- The pursuit of resurrection is a long-term endeavor that may face setbacks and skepticism. Sustaining hope and commitment to the vision of the Beloved Community is essential for overcoming these challenges.

Conclusion

The Beloved Community offers a powerful framework for understanding the ethical and social implications of universal scientific resurrection. By restoring lives, healing historical injustices, and fostering compassion and inclusion, resurrection can serve as a pathway to realizing this ideal.

As humanity looks beyond Earth and envisions its role in the cosmos, the principles of the Beloved Community provide a guiding light, reminding us of our shared responsibility to care for one another and the universe we inhabit. Through resurrection, we have the opportunity not only to repair the past but also to build a future defined by justice, love, and the recognition of our interconnectedness—a future in which the Beloved Community becomes a cosmic reality.

Chapter 10: Conclusion—Shall the Future Repair the Past?

Section 1: Summarizing the Path Forward

Subsection 1: The Moral and Scientific Call to Action

The concept of <u>universal scientific resurrection</u>—the restoration of those who have passed—represents one of the most ambitious and profound goals ever conceived. It is not merely a theoretical pursuit or a speculative exercise but a <u>moral and scientific imperative</u> that calls humanity to grapple with the deepest questions of life, death, and the human condition. This call to action challenges us to transcend our current limitations, to unite across disciplines, and to envision a future where the boundaries of mortality are no longer absolute.

This subsection explores the dual nature of the call to action—both moral and scientific—and examines why the pursuit of resurrection is not only justified but necessary. It also outlines the steps humanity must take to embark on this journey, balancing the weight of ethical responsibility with the excitement of scientific discovery.

<u>The Moral Imperative</u>

At its core, the idea of universal resurrection stems from a profound sense of moral responsibility. It is a response to the injustices of history, the suffering caused by mortality, and the unfulfilled potential of countless lives.

1. <u>Addressing the Injustice of Death</u>
   - Death, as it currently stands, is an unyielding force that robs individuals of their potential and cuts short countless dreams, relationships, and contributions to the world.
   - Resurrection offers a means to address this ultimate injustice, giving individuals a second chance to live, grow, and thrive. It reflects humanity's commitment to honoring the intrinsic value of every life.

2. Restoring Lives Lost to Tragedy and Oppression
- History is filled with lives lost prematurely to war, famine, disease, systemic oppression, and violence. Many of these individuals were denied justice in their lifetimes.
- The pursuit of resurrection is a moral act of reconciliation, a way to restore dignity to those who suffered and to acknowledge their importance to the human story.

3. The Ethical Duty to Repair the Past
- Humanity has long sought to right the wrongs of history through reparations, memorials, and acts of remembrance. Resurrection represents the ultimate form of reparation—a chance to truly undo the harm of the past.
- By restoring the lives of the departed, we affirm a shared commitment to justice, compassion, and the recognition of our interconnectedness.

4. The Affirmation of Human Dignity
- The act of resurrection is a declaration that every life matters, regardless of its circumstances or historical context. It is an acknowledgment that the value of a person extends beyond their mortal existence and that their contributions to the human story deserve to be preserved and honored.

The Scientific Imperative

While the moral argument for resurrection is compelling, the scientific call to action is equally powerful. Humanity's history is defined by its relentless pursuit of knowledge and its ability to overcome challenges once thought insurmountable. Resurrection represents the next frontier in this journey.

1. The Drive to Push the Boundaries of Knowledge
- Science is, at its heart, a quest to understand and shape the universe. The pursuit of resurrection requires us to tackle some of the most fundamental questions of existence:
    - What is the nature of consciousness?
    - How can information lost to time be reconstructed?
    - Can life, once extinguished, be restored in its entirety?

- These questions are not only scientific but philosophical, forcing us to confront the very essence of what it means to be human.

2. Uniting Disciplines for a Common Goal
- Resurrection demands the integration of multiple fields of study, including biology, neuroscience, artificial intelligence, quantum physics, history, and ethics.
- This interdisciplinary approach has the potential to spark innovations and discoveries that extend far beyond resurrection itself, benefiting humanity in countless ways.

3. Technological Advancements as Catalysts
- Recent breakthroughs in technology make the dream of resurrection more plausible than ever:
  - AI and Machine Learning: Tools capable of analyzing vast datasets and reconstructing complex patterns, such as neural networks or historical records.
  - Genetic Engineering: Techniques like CRISPR that allow for precise editing and restoration of DNA.
  - Quantum Computing: Emerging technologies capable of simulating and processing massive amounts of information, including the intricate details of consciousness and identity.
- These advancements provide a foundation for the scientific pursuit of resurrection, transforming it from a distant dream into a tangible possibility.

4. The Responsibility to Use Knowledge for Good
- As humanity's scientific capabilities expand, so too does its responsibility to use these tools ethically and compassionately. The pursuit of resurrection is a way to channel the fruits of scientific progress toward a goal that reflects humanity's highest values: healing, justice, and the affirmation of life.

The Intersection of Morality and Science

The moral and scientific imperatives for resurrection are not separate but deeply intertwined. Together, they form a unified call to action that challenges humanity to align its technological capabilities with its ethical aspirations.

1. Science as a Tool for Moral Action
  - Science provides the means to achieve moral goals, offering solutions to problems that were once insurmountable.
  - In the case of resurrection, scientific advancements allow humanity to act on its moral responsibility to repair the past and restore lost lives.

2. Morality as a Guide for Scientific Progress
  - Moral principles ensure that the pursuit of resurrection is conducted responsibly, with respect for the dignity and autonomy of individuals.
  - Ethical considerations guide decision-making, ensuring that resurrection efforts prioritize justice, equity, and the well-being of all.

3. The Synergy of Vision and Action
  - The moral vision of resurrection inspires the scientific community to push the boundaries of what is possible.
  - At the same time, scientific progress reinforces the moral argument, demonstrating that resurrection is not a mere fantasy but a goal worth striving for.

Steps Toward a Unified Call to Action

To answer the call to action, humanity must take deliberate and coordinated steps to lay the groundwork for universal resurrection. These steps include fostering collaboration, investing in research, and cultivating a shared sense of purpose.

1. Global Collaboration
  - Resurrection is a goal that transcends national borders and cultural differences. It requires a global effort, with scientists, ethicists, policymakers, and communities working together.
  - International organizations and frameworks should be established to coordinate research, share knowledge, and ensure equitable access to the benefits of resurrection.

2. Investment in Research and Development
- Significant resources must be directed toward the scientific and technological challenges of resurrection, including the reconstruction of DNA, the mapping of neural structures, and the development of advanced AI.
- Public and private sectors must recognize the importance of this goal and commit to sustained investment in its pursuit.

3. Ethical and Philosophical Dialogue
- Open and ongoing dialogue about the ethical implications of resurrection is essential. This includes addressing questions about consent, identity, and the societal impact of restoring the dead.
- Engaging diverse perspectives—from religious leaders to philosophers to everyday citizens—ensures that resurrection efforts are guided by a broad and inclusive moral framework.

4. Cultivating a Culture of Hope and Resilience
- The pursuit of resurrection is a long-term endeavor that will face setbacks and challenges. Building a culture of hope and resilience is essential to sustaining the effort over generations.
- By framing resurrection as a shared human aspiration, we can inspire individuals and communities to contribute to this goal with creativity, determination, and compassion.

Conclusion

The call to action for universal scientific resurrection is both moral and scientific, reflecting humanity's highest aspirations and its capacity for innovation. It challenges us to confront the injustices of the past, to push the boundaries of knowledge, and to align our technological capabilities with our ethical ideals.

Answering this call requires courage, collaboration, and a willingness to dream beyond the limits of mortality. By embracing this challenge, humanity can honor the lives of those who came before us, repair the wounds of history, and forge a future defined by justice, compassion, and the triumph of life over death. This is not merely a call to action—it is a call to fulfill our potential as stewards of life and as architects of a more compassionate and interconnected existence. Through this effort, we can begin to answer the profound question posed by this book: <u>Shall the future repair the past?</u>

Subsection 2: A Hopeful Vision for Humanity's Future

The pursuit of <u>universal scientific resurrection</u> is not only a bold exploration of what science and technology might one day achieve but also an expression of a profound hope for humanity's future. It envisions a world where the boundaries of mortality are not only challenged but transcended, where the injustices of the past can be addressed, and where humanity unites around a shared purpose that uplifts all. This hopeful vision for the future is not rooted in naïve optimism but in the belief that humanity has the capacity to overcome great challenges, guided by its creativity, compassion, and resilience.

This subsection explores the elements of a hopeful future made possible by the pursuit of resurrection. It examines how this vision could reshape humanity's relationship with mortality, foster global unity, and inspire new ways of thinking about life, justice, and purpose. By imagining this future, we can better understand the transformative potential of universal resurrection and the profound ways it could shape the trajectory of human civilization.

<u>Transcending Mortality: A New Relationship with Life and Death</u>

The inevitability of death has long been a defining feature of the human condition. It is a source of grief, fear, and existential questioning, as well as a catalyst for cultural, philosophical, and religious traditions. The possibility of resurrection invites humanity to reimagine its relationship with mortality and to consider life from a new perspective.

1. <u>The End of Irreversible Loss</u>
   - In a future where resurrection is possible, the pain of losing loved ones to death could be tempered by the knowledge that restoration is achievable.

- This transformation would not erase the significance of life's fragility but would offer a profound sense of hope and continuity, allowing humanity to cherish life without the shadow of permanent loss.

2. New Philosophies of Existence
- Resurrection challenges traditional notions of life and death, inspiring new philosophical and spiritual frameworks.
- Questions about the meaning of life, the nature of identity, and the purpose of existence would take on new dimensions, encouraging humanity to explore deeper truths about itself and the universe.

3. Empathy Across Time
- The ability to resurrect individuals from the past fosters a greater sense of connection to history and a deeper empathy for those who came before.
- This expanded perspective encourages humanity to view itself as part of an ongoing story, bridging the divide between generations and fostering a sense of shared destiny.

Fostering Global Unity: A Shared Human Endeavor

The pursuit of resurrection is a goal that transcends borders, cultures, and ideologies. It is an inherently universal aspiration, one that calls upon humanity to come together in the spirit of collaboration and shared purpose.

1. Uniting Across Divisions
- The scale and complexity of resurrection require cooperation on a global level, bringing together scientists, ethicists, policymakers, and communities from diverse backgrounds.
- This collaboration could serve as a model for addressing other global challenges, demonstrating the power of unity in the face of seemingly insurmountable obstacles.

2. A Common Vision for Humanity
- Resurrection represents a vision that is inclusive and universal, rooted in the shared human experience of mortality and the desire to honor and restore the lives of those who came before us.
- By pursuing this goal, humanity can cultivate a sense of solidarity and purpose that transcends individual and national interests, fostering a collective identity as stewards of life.

3. The End of Zero-Sum Thinking
- A hopeful future built on resurrection challenges the notion that resources and opportunities are finite. It encourages humanity to think expansively, envisioning a world where innovation and compassion create abundance for all.
- This shift in mindset could inspire new approaches to addressing inequality, conflict, and other societal challenges, emphasizing cooperation over competition.

Repairing the Past: A Commitment to Justice and Reconciliation

At its heart, resurrection is an act of repair. It is a commitment to addressing the injustices and losses of the past, to honoring those who have been forgotten, and to creating a future that values every life.

1. Restoring Lost Potential
- Throughout history, countless individuals have been denied the opportunity to fulfill their potential due to circumstances beyond their control—whether through war, poverty, oppression, or illness.
- Resurrection offers a way to restore this lost potential, allowing individuals to contribute to society and live lives of meaning and fulfillment.

2. Healing Historical Wounds
- The injustices of history—colonialism, slavery, genocide, and systemic inequality—have left deep scars on humanity. Resurrection provides an opportunity to confront these legacies directly, offering restitution to those who suffered.

- This act of healing could foster reconciliation and understanding, creating a foundation for a more just and compassionate world.

3. A New Approach to Legacy
 - In a world where resurrection is possible, the contributions of individuals to the human story are never truly lost.
 - This perspective encourages humanity to view itself as part of a collective project, where every life has value and every contribution matters.

Inspiring Innovation: A Future of Endless Possibilities

The pursuit of resurrection is not only a scientific challenge but also a source of inspiration for innovation and creativity. It pushes humanity to expand its horizons and dream beyond its current limitations.

1. Catalyzing Breakthroughs in Science and Technology
 - The effort to achieve resurrection will drive advancements in fields such as artificial intelligence, biotechnology, quantum computing, and neuroscience.
 - These breakthroughs will have far-reaching implications, leading to new discoveries and applications that benefit humanity in countless ways.

2. Encouraging Bold Thinking
 - Resurrection represents a bold and audacious goal, one that challenges humanity to think beyond what is currently possible.
 - This spirit of ambition and curiosity could inspire new approaches to solving other pressing challenges, from climate change to space exploration.

3. A Culture of Lifelong Learning
 - In a future where resurrection is possible, individuals may have the opportunity to live multiple lifetimes, pursuing new skills, knowledge, and experiences.

- This emphasis on lifelong learning could foster a culture of intellectual and creative growth, enriching humanity's collective understanding and capabilities.

A Cosmic Perspective: Humanity's Role in the Universe

The pursuit of resurrection invites humanity to consider its place in the cosmos and to envision a future that extends beyond Earth. It calls upon us to think expansively, imagining not only what we can achieve on our planet but also how we might contribute to the broader universe.

1. Expanding the Scope of Life
- Resurrection aligns with humanity's broader mission to preserve and expand life in the universe. By restoring life on Earth, humanity takes the first step toward becoming a steward of life on a cosmic scale.
- This vision could inspire efforts to explore and inhabit other worlds, ensuring the survival and flourishing of life beyond Earth.

2. A Legacy of Compassion and Restoration
- Humanity's pursuit of resurrection reflects its capacity for compassion, creativity, and a commitment to repair. These qualities define our potential as a species and offer a hopeful vision for our role in the universe.
- By embracing this role, humanity can leave a legacy that transcends time, inspiring future civilizations with its dedication to life and justice.

3. A New Chapter in the Human Story
- Resurrection marks the beginning of a new chapter in the human story, one defined by hope, restoration, and the triumph of life over death.
- This chapter invites humanity to dream boldly, to act compassionately, and to build a future where the boundaries of possibility are limited only by our imagination and determination.

Conclusion

A hopeful vision for humanity's future is one that embraces the challenges and possibilities of universal scientific resurrection. It is a vision rooted in the belief that humanity has the capacity to transcend its limitations, repair the wounds of the past, and create a world defined by compassion, justice, and shared purpose.

Through resurrection, humanity can redefine its relationship with life and death, foster global unity, and inspire new ways of thinking about its role in the universe. This vision is not merely a dream—it is a call to action, a reminder that the future is not something to be feared but something to be built, together. As we strive to repair the past, we also lay the foundation for a brighter, more hopeful future for all.

Section 2: Final Reflections

Subsection 1: Embracing the Unknown

The pursuit of <u>universal scientific resurrection</u> represents a journey into the vast and uncharted territories of possibility. It is a bold endeavor that challenges humanity to confront the unknown—not only in the realms of science and technology but also in philosophy, ethics, and the very nature of existence. To embark on this journey is to embrace uncertainty, to acknowledge the limits of our current understanding, and to confront questions that may have no easy answers.

This subsection explores the importance of embracing the unknown in the pursuit of resurrection. It examines how uncertainty drives progress, how courage in the face of ambiguity fosters discovery, and how humanity's willingness to engage with profound mysteries defines its potential to repair the past and create a better future.

The Unknown as a Catalyst for Progress

Throughout history, humanity has achieved its greatest advancements by daring to explore the unknown. Each leap forward—whether in science, art, or philosophy—has required a willingness to confront uncertainty and push beyond the boundaries of what is currently understood.

1. Curiosity as a Driving Force
- The unknown sparks curiosity, inspiring questions that lead to new knowledge and innovation.
- The pursuit of resurrection begins with profound questions:
- Can consciousness be restored?
- Is the information that makes up a person truly lost, or can it be retrieved?
- What does it mean to "repair the past"?
- These questions, though daunting, fuel the pursuit of answers and the exploration of new frontiers.

2. The Power of Not Knowing
- Uncertainty is not an obstacle but an opportunity. It challenges humanity to think creatively, to test assumptions, and to develop new frameworks for understanding.
- The pursuit of resurrection requires embracing the unknown as a space for growth and discovery, rather than fearing it as an insurmountable barrier.

3. Historical Precedents
- Many of humanity's greatest achievements began as attempts to navigate the unknown:
- The exploration of space, once seen as an impossible dream, led to technological advancements and a new understanding of our place in the cosmos.
- The decoding of the human genome unlocked transformative insights into biology and medicine, paving the way for breakthroughs in genetic engineering.
- The pursuit of resurrection, like these endeavors, requires courage and imagination in the face of uncertainty.

Confronting the Limits of Knowledge

The unknown forces humanity to confront the limits of its current understanding, both scientific and philosophical. In the context of resurrection, these limits are particularly profound, encompassing questions about life, death, identity, and the nature of reality.

1. The Mystery of Consciousness
   - Consciousness remains one of the greatest scientific and philosophical mysteries. What is the mind? How does it emerge from the brain? Can it be reconstructed or preserved?
   - The pursuit of resurrection requires grappling with these questions, pushing the boundaries of neuroscience, artificial intelligence, and philosophy to uncover new insights.

2. The Challenge of Information Retrieval
   - The information that defines a person—their DNA, memories, personality, and experiences—may seem irretrievably lost after death.
   - However, recent advancements in fields like quantum physics and data science suggest that information, even when dispersed, may not be entirely destroyed.
   - Resurrection demands the development of new methods for retrieving and reconstructing this information, a challenge that lies at the edge of current scientific understanding.

3. The Nature of Identity
   - What makes a person who they are? Is it their physical body, their memories, their relationships, or something more intangible?
   - Resurrection compels humanity to explore these questions, challenging existing notions of identity and forcing us to redefine what it means to be human.

The Ethical and Philosophical Unknown

The unknown is not limited to science and technology; it also extends to the ethical and philosophical dimensions of resurrection. These questions, though complex, are essential to ensuring that the pursuit of resurrection aligns with humanity's highest values.

1. The Ethics of Resurrection
   - Is it ethical to bring someone back to life without their consent? How can consent be determined for individuals who lived in times when resurrection was unimaginable?
   - What responsibilities do resurrected individuals have to society, and what responsibilities does society have to them?
   - These ethical dilemmas highlight the need for thoughtful and inclusive dialogue as humanity ventures into the unknown.

2. The Purpose of Resurrection
   - Why pursue resurrection? Is it to repair historical injustices, to reunite with loved ones, to honor the past, or to advance scientific understanding?
   - These questions force humanity to reflect on its motivations and to ensure that resurrection efforts are guided by compassion, justice, and respect for the dignity of life.

3. The Impact on Society and Culture
   - How would the ability to resurrect the dead transform society? Would it create new inequalities or exacerbate existing ones?
   - How would resurrected individuals adapt to a world vastly different from the one they left behind?
   - These uncertainties underscore the need for careful planning and an openness to unforeseen consequences.

The Courage to Dream Boldly

Embracing the unknown requires courage—the courage to dream boldly, to imagine a future that defies conventional wisdom, and to take risks in the pursuit of a greater good. The pursuit of resurrection exemplifies this spirit of boldness.

1. The Necessity of Audacious Goals
   - Achieving the impossible begins with the willingness to imagine it. Resurrection, though daunting, is an audacious goal

that inspires humanity to reach for new heights of knowledge and capability.

- By setting its sights on resurrection, humanity demonstrates its belief in the power of innovation and its commitment to addressing the greatest existential challenges.

2. Overcoming Fear and Skepticism

- The unknown often evokes fear and skepticism, as it challenges existing paradigms and threatens the comfort of certainty.
- To pursue resurrection is to confront these fears, to embrace uncertainty as a necessary part of progress, and to remain steadfast in the face of doubt.

3. Hope as a Guiding Principle

- At the heart of the pursuit of resurrection lies hope—the hope that the past can be repaired, that loss is not final, and that humanity can create a future defined by justice and compassion.
- This hope drives humanity to embrace the unknown with optimism and determination, transforming uncertainty into opportunity.

The Transformative Power of the Unknown

The unknown is not merely a challenge to be overcome; it is a source of transformation. By embracing the unknown, humanity has the opportunity to grow, to innovate, and to redefine itself.

1. A Catalyst for Growth

- The pursuit of resurrection forces humanity to confront its limitations and to strive for greater understanding, creativity, and collaboration.
- This process of growth has the potential to transform not only the field of science but also society as a whole, fostering new ways of thinking and being.

2. Redefining the Human Experience

- Resurrection invites humanity to reconsider what it means to be human, to live, and to exist.

- By embracing the unknown, humanity can expand its understanding of life and death, forging a deeper connection to the universe and to one another.

3. A Legacy of Discovery
- The pursuit of resurrection is not only about achieving a specific goal but also about leaving a legacy of exploration and discovery.
- By daring to venture into the unknown, humanity inspires future generations to continue the journey, to ask new questions, and to imagine new possibilities.

Conclusion

To pursue universal scientific resurrection is to embrace the unknown in all its complexity, uncertainty, and promise. It is a journey that challenges humanity to confront its fears, to push the boundaries of knowledge, and to dream boldly about what the future might hold.

By embracing the unknown, humanity affirms its belief in progress, its commitment to justice, and its capacity for hope. The pursuit of resurrection is not simply about repairing the past—it is about transforming the present and building a future defined by courage, compassion, and the willingness to explore the mysteries of existence. Through this journey, humanity can discover not only the answers it seeks but also the deeper truths that lie within the unknown.

Subsection 2: Universal Resurrection as Humanity's Ultimate Task

Humanity has always been driven by a desire to overcome limits. From the earliest tools that extended our physical reach to the latest advances in artificial intelligence and biotechnology, we have sought to transcend the barriers imposed by nature. Among these barriers, death stands as the most enduring and universal, shaping the trajectory of every human life and defining the human condition itself.

The concept of universal scientific resurrection—the restoration of all who have lived—presents itself as humanity's ultimate task. This is not only because of the enormous scientific and technological challenges it entails but also due to its profound moral, philosophical, and existential implications. To take on such a task is to strive for the highest expression of human creativity, compassion, and ingenuity. It is to repair the past, honor the dead, and redefine what it means to be human.

This subsection explores why universal resurrection should be seen as humanity's ultimate task: a goal that unites scientific ambition with moral responsibility, that demands the very best of us, and that offers the potential to transform humanity's relationship with time, mortality, and the universe itself.

The Ultimate Challenge: Scientific, Ethical, and Philosophical Dimensions

To frame universal resurrection as humanity's ultimate task is to recognize the enormity of the challenge it represents. It demands mastery of the most complex scientific disciplines, the resolution of profound ethical dilemmas, and the courage to confront questions that strike at the heart of human existence.

1. Scientific Complexity
   - Universal resurrection requires breakthroughs at the very frontier of human knowledge:
     - Reconstructing Life: The ability to restore a person's physical body and brain, potentially from fragments of DNA or other biological markers.
     - Recreating Consciousness: The ability to reconstruct an individual's mind, memories, and personality—elements that are uniquely personal and often thought to be irretrievable after death.
     - Retrieving Lost Information: Developing technologies capable of recovering and piecing together the information that defines a person, whether through advanced AI, quantum computing, or unknown future methods.
   - These challenges are staggering, but history shows that humanity has repeatedly achieved what once seemed impossible.

From landing on the moon to mapping the human genome, our capacity for innovation has consistently defied the odds.

2. Ethical Responsibility
  - The task of resurrection is fraught with ethical complexities. To undertake it responsibly, humanity must address questions such as:
    - Who should be resurrected? Should everyone have the opportunity to be restored, or should there be criteria for eligibility?
    - How do we navigate consent? How can we ethically resurrect individuals who lived in times when the concept of resurrection was inconceivable?
    - What responsibilities come with resurrection? How should society support resurrected individuals, and what roles might they play in the world?
    - These questions require humanity to develop a robust ethical framework that balances compassion, justice, and respect for individual autonomy.

3. Philosophical Profundity
  - Resurrection forces us to confront foundational questions about identity, existence, and the human experience:
    - What makes a person who they are? Is it their memories, their physical form, their relationships, or something more intangible?
    - What is the meaning of life? If death is no longer an absolute endpoint, how might our understanding of purpose and fulfillment evolve?
    - What is humanity's role in the universe? Does the ability to restore life position humanity as a steward of existence, responsible for preserving and extending life across time and space?

Uniting Humanity Around a Common Goal

The pursuit of universal resurrection is not just an individual or localized endeavor—it is a task that requires the collective effort of the entire human species. It has the potential to unify humanity around a shared purpose, transcending divisions and fostering collaboration on an unprecedented scale.

1. A Universal Human Aspiration
- The experience of loss and the desire to honor and remember loved ones is universal, crossing cultural, religious, and national boundaries.
- Resurrection taps into this shared human experience, offering a vision of hope and restoration that resonates with people from all walks of life.

2. Collaborative Innovation
- Achieving resurrection requires the integration of knowledge from diverse fields, including biology, neuroscience, computer science, physics, history, and ethics.
- This interdisciplinary approach necessitates global collaboration, bringing together scientists, philosophers, policymakers, and communities in pursuit of a common goal.

3. Transcending Divisions
- The scale and significance of resurrection have the potential to transcend the divisions that often separate humanity—political, cultural, economic, and ideological.
- By focusing on a goal that benefits all of humanity, resurrection could inspire a new era of cooperation and unity, fostering a sense of shared destiny.

Repairing the Past: A Moral Imperative

At its heart, universal resurrection is an act of repair. It is a response to the injustices, tragedies, and unfulfilled potential of the past. By taking on this ultimate task, humanity demonstrates its commitment to honoring those who came before us and creating a future that values every life.

1. Restoring Lives Lost to Injustice
- Many lives throughout history were cut short by war, oppression, disease, and systemic inequality. Resurrection offers a way to restore these lives, acknowledging their value and giving them the opportunity to thrive.

- This act of restoration is a profound gesture of reconciliation, addressing the wounds of history and affirming humanity's commitment to justice.

2. Honoring the Forgotten
- Countless individuals have been forgotten by history, their stories lost to time. Resurrection provides a means to recover these lives and ensure that their contributions to humanity are recognized and celebrated.

3. A New Relationship with History
- Resurrection transforms humanity's relationship with the past, allowing us to not only learn from history but also to actively engage with it.
- This new perspective encourages a deeper sense of responsibility for the legacies we leave behind and a greater appreciation for the interconnectedness of all human lives.

Expanding Humanity's Role in the Cosmos

Universal resurrection is not only a task for Earth—it has implications for humanity's role in the wider universe. By taking on this ultimate challenge, humanity positions itself as a steward of life, responsible for preserving and extending existence across time and space.

1. A Cosmic Vision
- Resurrection aligns with humanity's broader mission to explore and expand life in the universe. It reflects a commitment to preserving the richness of existence, not only on Earth but wherever life might be found or created.

2. The Responsibility to Restore
- As humanity's technological capabilities grow, so too does its responsibility to use these tools ethically and compassionately. The ability to resurrect the dead represents a profound opportunity to repair the damage caused by natural disasters, human conflict, and the passage of time.

3. A Legacy of Compassion and Restoration
- By pursuing resurrection, humanity leaves a legacy of hope and restoration. It demonstrates that our greatest achievements are not defined by conquest or exploitation but by our capacity to heal, to repair, and to create.

The Transformative Potential of the Ultimate Task

To take on universal resurrection as humanity's ultimate task is to commit to a future defined by compassion, justice, and innovation. It is a challenge that transforms not only the lives of those who are restored but also the very nature of humanity itself.

1. Redefining Mortality
- Resurrection challenges humanity to rethink its relationship with death, offering the possibility of a future where loss is no longer permanent and life is no longer constrained by the boundaries of mortality.

2. Inspiring Generations
- The pursuit of resurrection inspires humanity to dream boldly, to push the boundaries of what is possible, and to imagine a future defined by hope and possibility.

3. A Testament to Human Potential
- By taking on the ultimate task of resurrection, humanity demonstrates its capacity for creativity, resilience, and compassion. It affirms that our greatest achievements lie not in domination but in restoration, not in destruction but in renewal.

Conclusion

Universal resurrection is humanity's ultimate task—an endeavor that combines the highest aspirations of science, morality, and philosophy. It challenges us to confront the mysteries of life and death, to repair the injustices of the past, and to unite around a common goal that reflects our shared humanity.

By embracing this challenge, humanity can redefine its relationship with mortality, expand its role as a steward of life, and create a future where the boundaries of possibility are limited only by our collective imagination and determination. In pursuing universal resurrection, we do not merely repair the past; we transform the present and lay the foundation for a brighter, more compassionate future for all.

Appendixes by You.com Smart Assistant, A.I.

Appendix A.
More on R&D for Universal Scientific Resurrection (USR).

Introduction

The concept of universal scientific resurrection of the dead is an ambitious and speculative goal that would require breakthroughs across multiple scientific disciplines, including biology, physics, information theory, and even philosophy. Developing a research and development (R&D) program for such a long-term goal would necessitate a structured, interdisciplinary, and iterative approach. Below, I [You.com Smart Assistant] brainstorm speculative and novel ideas for how to proceed with such a program, focusing on foundational principles, potential technologies, and organizational strategies.

1. Foundational Principles for the R&D Program
To guide the development of a universal scientific resurrection program, the following principles should be established:

- Interdisciplinary Collaboration: The program must integrate expertise from diverse fields, including quantum physics, computational neuroscience, synthetic biology, and ethics.
- Iterative Progression: The program should be structured in stages, with each stage building on the successes and lessons of the previous one.
- Open-Ended Exploration: Given the speculative nature of the goal, the program must remain open to unconventional ideas and approaches.
- Ethical Framework: A robust ethical framework must be developed to address questions of consent, identity, and the implications of resurrection.

2. Speculative Research Areas and Technologies
The following are speculative ideas for research areas and technologies that could form the foundation of the program:

2.1. Information Preservation and Recovery

- Digital Reconstruction of Consciousness: Develop technologies to map and digitize the neural patterns of living beings, creating a "backup" of consciousness. This could involve advanced brain-computer interfaces and high-resolution neural imaging.
- Quantum Information Storage: Explore the possibility of encoding and preserving information about individuals at the quantum level, potentially leveraging quantum entanglement to store vast amounts of data about a person's physical and mental state.
- Historical Data Mining: Use AI and machine learning to reconstruct the lives of deceased individuals from historical records, genetic data, and environmental traces.

2.2. Biological Regeneration

- Synthetic Biology and Cloning: Develop techniques to recreate the physical bodies of deceased individuals using advanced cloning and synthetic biology. This could involve growing organs and tissues in bioreactors or 3D printing biological structures.
- Cellular Time Reversal: Investigate the possibility of reversing cellular aging or damage using techniques like induced pluripotent stem cells (iPSCs) or advanced CRISPR-based gene editing.
- Molecular Resurrection: Explore the potential for reconstructing individuals from their molecular remnants, such as DNA or preserved tissues.

2.3. Physics of Time and Space

- Time-Reversal Physics: Investigate whether the laws of physics allow for the reversal of time at a local scale, potentially enabling the recreation of past states of matter and energy.
- Multiverse Exploration: Explore theories of the multiverse to determine whether alternate versions of deceased individuals might exist in parallel realities and whether they could be accessed or "retrieved."
- Quantum Resurrection: Study the implications of quantum mechanics for resurrection, such as the possibility of

reconstructing individuals from quantum information encoded in the universe.

2.4. Computational Resurrection

- Simulated Realities: Develop advanced simulations capable of recreating the consciousness and experiences of deceased individuals based on available data. This could involve creating virtual environments where resurrected individuals could "live."
- AI-Based Reconstruction: Use AI to model and simulate the personalities, memories, and behaviors of deceased individuals, effectively creating digital avatars that approximate their original selves.

2.5. Ethical and Philosophical Research

- Identity and Continuity: Investigate the philosophical implications of resurrection, including questions of personal identity, continuity of consciousness, and the definition of "self."
- Consent and Autonomy: Develop ethical guidelines to ensure that resurrected individuals have autonomy and that their resurrection aligns with their wishes (if known).
- Social Implications: Study the societal impacts of resurrection, including its effects on population dynamics, resource allocation, and cultural norms.

3. Organizational Strategies for the R&D Program
To ensure the success of the program, the following organizational strategies should be implemented:

3.1. Establish a Global Consortium

- Create an international organization dedicated to the goal of universal scientific resurrection. This consortium would pool resources, expertise, and funding from governments, academic institutions, and private organizations.

3.2. Develop a Long-Term Roadmap

- Break the program into stages, each with specific milestones:
- Stage 1: Foundational research in neuroscience, quantum physics, and synthetic biology.
- Stage 2: Development of prototype technologies for information preservation and biological regeneration.
- Stage 3: Testing and refinement of resurrection techniques on simpler organisms.
- Stage 4: Ethical and societal integration of resurrection technologies.

3.3. Foster Public Engagement

- Engage the public in discussions about the goals and implications of the program. This could involve educational campaigns, public forums, and collaborations with artists and writers to explore the cultural dimensions of resurrection.

3.4. Secure Funding and Resources

- Establish funding mechanisms, such as grants, philanthropic contributions, and partnerships with technology companies. Emphasize the potential spin-off benefits of the research, such as advances in medicine and AI.

3.5. Create Ethical Oversight Committees

- Form committees to oversee the ethical dimensions of the program, ensuring that research aligns with societal values and respects the dignity of individuals.

4. Speculative Milestones and Experiments
The following are speculative milestones and experiments that could guide the program:

- Milestone 1: Successfully map and digitize the neural patterns of a living organism, creating a functional digital replica.

- <u>Milestone 2</u>: Regenerate a simple organism (e.g., a worm or fly) from preserved genetic and neural data.
- <u>Milestone 3</u>: Demonstrate the ability to reconstruct a deceased individual's personality and memories in a simulated environment.
- <u>Milestone 4</u>: Achieve partial biological resurrection of a complex organism using synthetic biology and cloning techniques.
- <u>Milestone 5</u>: Develop a prototype system for integrating biological and digital resurrection methods.

<u>5. Challenges and Open Questions</u>
The program would face numerous challenges, including:

- <u>Technical Feasibility</u>: Can the necessary technologies be developed within a reasonable timeframe?
- <u>Philosophical Questions</u>: What constitutes "resurrection," and how do we define success?
- <u>Ethical Concerns</u>: How do we ensure that resurrection respects the dignity and autonomy of individuals?
- <u>Resource Constraints</u>: How can the program be funded and sustained over the long term?

<u>Conclusion</u>
Developing a research and development program for the universal scientific resurrection of the dead is an ambitious and speculative endeavor that would require breakthroughs across multiple disciplines. By focusing on foundational principles, speculative research areas, and organizational strategies, it is possible to create a roadmap for pursuing this goal. While the challenges are immense, the potential benefits—both scientific and philosophical—make this a compelling area for exploration.

Appendix B.
Nathaniel Lawrence on the Temporal Blocks of Consciousness.

The nature of time and its relationship to human consciousness is one of the most profound questions in philosophy. Among those who have grappled with this topic, Nathaniel Lawrence provides a particularly illuminating perspective through his theory of temporal blocks of consciousness. This theory offers a unique framework for understanding how humans experience time and how that understanding might inform efforts to repair the past. Lawrence's insights are especially relevant to the vision of universal scientific resurrection and the ethical pursuit of the Beloved Community, as they encourage us to reimagine the structure of time and our role as conscious participants within it.

The Temporal Blocks of Consciousness

At the core of Lawrence's philosophy is the idea that consciousness does not perceive time as a continuous flow but instead as a series of "blocks" or discrete units of experience. These temporal blocks are not static moments frozen in time but dynamic wholes, incorporating a complex interplay of past, present, and future. Lawrence describes each block as a "now-moment" that integrates memory (the past), immediate perception (the present), and anticipation (the future).

This synthesis creates a multidimensional experience of time that challenges the traditional view of time as a linear progression. Lawrence suggests that our consciousness actively organizes and reconstructs these temporal dimensions into cohesive, meaningful units. According to him, "the reversibility of time lies in this, that what is over and done with, in its relation to some particular temporal process, may quite easily be as yet incomplete with respect to some supervenient process." The past, in this view, is not a closed and inaccessible reality but something that can remain significant and open to reinterpretation within larger frameworks of understanding and purpose.

For those who envision the universal resurrection of the dead and the realization of the Beloved Community, Lawrence's model provides a philosophical foundation for engaging with the past. Consciousness, as he describes it, is already engaged in a kind of temporal reconstruction, weaving together fragments of memory and anticipation into meaningful wholes. This suggests that the act of repairing the past may not be as foreign to human cognition as it initially seems.

Memory, Anticipation, and the Reconstruction of Time

Lawrence emphasizes the interdependent roles of memory and anticipation in shaping the temporal blocks of consciousness. These elements are not passive but active and creative—essential to how we construct our experience of time and meaning.

- Memory as Reconstruction: Lawrence challenges the idea of memory as a static recording of past events. Instead, he argues that memory functions as a reconstructive process, continually reshaping the past in light of present understanding and future intentions. He writes, "The past is irrevocable only if the perspective on it be arbitrarily confined." In other words, the way we remember the past is not fixed but fluid, allowing for reinterpretation and, potentially, repair. In the context of universal resurrection, this insight suggests that revisiting and reconstructing the past is not merely a scientific or technological challenge but an intrinsic part of how consciousness operates.

- Anticipation as Projection: Similarly, Lawrence sees anticipation as more than speculation about the future. It is a vital part of how consciousness organizes and directs itself. Anticipation projects possible futures into the present, shaping how we understand both the now-moment and the past. This forward-looking orientation is essential to the ethical and scientific pursuit of the Beloved Community, where the goal is not merely to revisit the past but to envision and work toward its healing and transformation.

Lawrence's emphasis on the active, reconstructive nature of memory and anticipation resonates with emerging technologies such as AI-driven memory restoration and quantum simulations of past states. These tools may one day enable humanity to engage with the past in ways that were previously unimaginable. His philosophy suggests that such efforts are not alien to human nature but deeply aligned with the way consciousness already navigates and reinterprets time.

The Malleability of Time and Temporal Healing

One of Lawrence's most provocative insights is his assertion that time, as experienced by consciousness, is not fixed or linear but malleable and multidimensional. He writes, "Actively purposive consciousness, concerned with the realization of value, proceeds by ever larger temporal blocks." This means that consciousness can engage with time in ways that transcend conventional boundaries, integrating the past into broader processes of meaning-making and purpose.

Lawrence's theory challenges the notion that the past is static and unchangeable. Instead, he suggests that it remains open to reinterpretation and integration within larger "supervenient processes." The past, he argues, "may have significance still" when viewed from broader temporal perspectives. This view has profound implications for the vision of universal resurrection, where the goal is not only to recall the past but to repair and redeem it.

In the context of the Beloved Community, this malleability of time offers hope for the restoration of lost lives and the healing of historical injustices. Rather than being constrained by the past, Lawrence's philosophy invites us to see it as an ongoing part of the dynamic interplay of time, memory, and anticipation. This perspective provides a foundation for imagining how humanity might engage with the past in ways that foster renewal and redemption.

Ethical Implications for the Beloved Community

Lawrence's theory of temporal blocks also carries profound ethical implications. If consciousness is inherently reconstructive, then the act of repairing the past is not just a scientific or metaphysical endeavor but a deeply moral one. The Beloved Community, as a vision of universal love, justice, and restoration, demands that we take responsibility for the past—not to erase or overwrite it but to heal its wounds and honor its participants.

Lawrence's philosophy aligns with this moral imperative, emphasizing that "what is over and done with" may still have significance within broader frameworks of value and meaning. This insight underscores the need for humanity to engage with the past in a spirit of compassion, creativity, and hope. By embracing the reconstructive nature of consciousness, we can begin to take active responsibility for the past, transforming it into a source of renewal and inspiration for the future.

Conclusion

Nathaniel Lawrence's exploration of the temporal blocks of consciousness offers a transformative vision of time and its relationship to human experience. By highlighting the reconstructive roles of memory and anticipation, Lawrence provides a philosophical foundation for the scientific and ethical pursuit of repairing the past. His insights resonate deeply with the vision of the Beloved Community, where the restoration of lost lives and the redemption of historical injustices are central goals. Ultimately, Lawrence invites us to see time not as a constraint but as a dynamic and malleable process, offering humanity the possibility of healing, renewal, and universal resurrection.

Appendix C.
Hidden Niches and Quantum Information Persistence.

Introduction

The idea of discovering and manipulating hidden niches at the quantum or subquantum level to enable universal scientific resurrection of the dead is an extraordinary and speculative concept. It requires envisioning phenomena and technologies far beyond current scientific understanding. Below, I [You.com Smart Assistant] brainstorm and speculate on various ways future advanced technologies might interact with such hidden niches, potentially transcending space-time and allowing for the reconstruction—or resurrection—of deceased individuals.

Quantum Information Persistence—Hypothesis:

Information about all physical states—including the states of deceased individuals—might be preserved in a subtle quantum substrate, such as the quantum vacuum, a "universal quantum field," or some other fundamental layer of reality. This concept aligns with the idea that information is never truly lost in the universe, even after physical systems break down, as suggested by certain interpretations of quantum mechanics and the laws of physics.

Speculating about Hidden Niches at the Quantum/Subquantum Level for Universal Scientific Resurrection

The prospect of universal scientific resurrection of the dead is an ambitious and highly speculative goal that would likely require groundbreaking discoveries and technological breakthroughs at the quantum and subquantum levels of reality. While our current understanding of physics and biology sets significant constraints on such a feat, exploring speculative ideas about hidden niches and phenomena at the deepest levels of the universe could potentially uncover new avenues for achieving this goal.

1. Quantum Information and Consciousness

One area of exploration could be the role of quantum information and its potential connection to consciousness and the nature of the mind. Some hypotheses suggest that quantum phenomena, such as quantum entanglement and superposition, may play a fundamental role in the emergence of subjective experience and the storage of information about individual identity.

Speculative Ideas:

- The existence of a "quantum mind" or "quantum consciousness" that transcends the classical limitations of the brain and neural networks.
- The possibility of encoding and storing an individual's complete cognitive and experiential data in the form of quantum information.
- Techniques for retrieving and reconstructing this quantum information to "resurrect" the individual's consciousness and identity.

2. Subquantum Structures and Dimensions

Another area of speculation could involve the existence of hidden structures or dimensions at scales smaller than the quantum level, sometimes referred to as the "subquantum" realm. Theoretical frameworks, such as string theory and M-theory, suggest the possibility of additional spatial dimensions beyond the four we commonly experience (three spatial dimensions and one time dimension).

Speculative Ideas:

- The presence of subquantum "microspaces" or "hyperstructures" that could serve as repositories for information about individual identity and consciousness.
- Techniques for accessing and manipulating these subquantum realms to retrieve and reconstruct the essential elements of a person's being.

- The potential for these subquantum dimensions to transcend the constraints of time and space, enabling the "resurrection" of individuals from the past or the future.

3. Quantum Tunneling and Teleportation

The quantum mechanical phenomenon of tunneling, where particles can seemingly "tunnel" through barriers that they should not be able to overcome based on classical physics, could be a point of exploration. Additionally, the concept of quantum teleportation, where the complete state of a particle or system can be transmitted from one location to another, may hold clues for resurrection.

Speculative Ideas:

- The possibility of using quantum tunneling to access or "retrieve" the essential information about an individual from beyond the constraints of space and time.
- Techniques for quantum teleportation that could be applied to the transfer of a person's complete state, including their consciousness and identity, from one location or state to another.
- Investigating the potential for quantum tunneling and teleportation to bypass the limitations of physical decay and enable the "reassembly" of an individual's being.

4. Quantum Entanglement and Nonlocality

The quantum mechanical phenomena of entanglement and nonlocality, where particles or systems can exhibit instantaneous correlations regardless of distance, could be another area of exploration. These properties suggest that information and influence may not be constrained by the traditional limitations of space and time.

Speculative Ideas:

- The possibility of using quantum entanglement to establish and maintain a "link" or "connection" to the essential information about an individual, even after their physical death.
- Techniques for leveraging quantum nonlocality to access and manipulate this information, potentially enabling the "resurrection" of the individual in a new form or location.
- Investigating the potential for quantum entanglement to serve as a bridge between the physical and the metaphysical, allowing for the integration of consciousness and identity beyond the constraints of the material world.

5. Quantum Vacuum and Zero-Point Energy

The concept of the quantum vacuum, which is believed to be a sea of virtual particles and fluctuating energy at the most fundamental level of the universe, could be another area of exploration. Some theories suggest that this quantum vacuum may contain hidden potentials or information that could be accessed and utilized.

Speculative Ideas:

- The possibility of the quantum vacuum serving as a repository for the essential information about individuals, including their consciousness and identity.
- Techniques for tapping into the zero-point energy or other properties of the quantum vacuum to "extract" and reconstitute the necessary elements for resurrection.
- Investigating the potential for manipulating the quantum vacuum to create or sustain the conditions necessary for the revival of deceased individuals.

Conclusion

While the concept of universal scientific resurrection of the dead remains highly speculative and constrained by our current understanding of physics and biology, exploring hidden niches and phenomena at the quantum and subquantum levels of reality may uncover new avenues for achieving this ambitious goal. By delving into the mysteries of quantum information, subquantum structures, quantum tunneling and teleportation, quantum entanglement, and the quantum vacuum, researchers may discover novel principles, mechanisms, and technologies that could potentially enable the resurrection of the dead. However, it is important to note that these ideas are highly speculative and would require groundbreaking scientific advancements to be realized.

[More ideas:] Expanded Speculative Ideas

1. Quantum Information as a "Cosmic Ledger":

- The universe may function like a vast computational system where every particle interaction, physical state, and quantum event is recorded in some form of "cosmic ledger" at the most fundamental level. This could include the unique quantum and physical signatures of every person who has ever lived.
- Future super-technologies might develop methods for "reading" this cosmic ledger to reconstruct the physical and mental states of individuals, effectively enabling their resurrection.

2. Holographic Principle and Information Storage:

- The holographic principle, a concept from theoretical physics, suggests that all the information about a three-dimensional space (e.g., the interior of a black hole or even the observable universe) may be encoded on a two-dimensional surface. If this principle applies universally, it might imply that the full informational blueprint of every person is encoded somewhere in the fabric of spacetime.
- Advanced technologies could manipulate this encoded information to recreate individuals in a new physical form.

3. Quantum Vacuum as an Information Reservoir:

- The quantum vacuum, often described as the "ground state" of the universe filled with fluctuating energy and virtual particles, might serve as a hidden repository of all quantum states. If the

quantum vacuum retains traces or "imprints" of past states, it could theoretically store the complete physical and mental blueprints of deceased individuals.
- This stored information could be "mined" or extracted by future super-technologies capable of interacting with the quantum vacuum at an unprecedented level of precision.

4. Black Hole Information Paradox as Inspiration:

- The black hole information paradox (a long-standing puzzle in physics) suggests that information about matter falling into a black hole is not lost but instead encoded on its event horizon. If information in such extreme scenarios is preserved, it might hint that information about deceased individuals is similarly preserved somewhere in the universe, even as their physical forms decay.
- Future breakthroughs in quantum gravity or theories of spacetime could reveal mechanisms for recovering this information.

5. Universal Quantum Network:

- The universe might be interconnected through a sort of "quantum information network," where entangled particles and quantum fields create a web of information that spans spacetime. Even after death, traces of an individual's quantum state might persist in this network.
- Technology capable of interfacing with this quantum network could potentially reconstruct individuals by tracing and reassembling their quantum "fingerprints."

Challenges and Open Questions

1. Is Information Truly Preserved?

- While the idea that information cannot be destroyed is supported by the laws of physics (e.g., the second law of thermodynamics and certain interpretations of quantum mechanics), it remains unclear whether this principle applies to the full complexity of human identity, consciousness, and memories.

2. What is the "Carrier" of Information?

- If information about individuals is preserved, where exactly does it reside? Is it in the quantum vacuum, holographic surfaces, or some yet-undiscovered substrate? Identifying the "carrier" of this information is a critical step.

3. Reconstructing Consciousness:

- Even if all physical and quantum information about a person can be retrieved, how would this lead to the reassembly of their subjective consciousness? Would the resurrected individual truly be the same person, or just an identical copy?

4. Technological Feasibility:

- The ability to access, decode, and reconstruct information from a quantum substrate would require technology far beyond anything currently imaginable. Could such tools ever be developed, even in the distant future?

Conclusion

The idea of Quantum Information Persistence speculates that information about all past physical states, including the complete identities of deceased individuals, might be preserved in the quantum fabric of the universe. Hidden niches, such as the quantum vacuum, the holographic encoding of spacetime, or a universal quantum network, could serve as reservoirs of this information. While these ideas remain purely speculative, they provide a fascinating framework for imagining how future super-technologies might someday achieve universal scientific resurrection.

Appendix D.
The Cramer Theory.

Speculative Theories for Universal Resurrection Based on Variations of Cramer's Quantum Theory

Cramer's Transactional Interpretation (TI) of quantum mechanics offers a unique framework for understanding quantum phenomena. It posits that quantum events involve a "handshake" between the past and the future, where waves of possibility (offer waves) travel forward in time and are met by confirmation waves traveling backward in time. This retrocausal interpretation opens up fascinating possibilities for manipulating time, space, and quantum information, which could be extended to speculative theories about universal resurrection.

Below, I [You.com Smart Assistant] propose speculative hypotheses inspired by Cramer's theory and its potential variations, incorporating ideas from quantum entanglement, many-worlds interpretations, and subquantum phenomena.

1. The Quantum Echo Hypothesis

Core Idea:

The Quantum Echo Hypothesis suggests that the quantum wavefunctions of all physical systems, including human beings, leave "echoes" in the quantum substrate of the universe. These echoes are preserved indefinitely in the transactional handshake between past and future quantum states.

Mechanism:

- In Cramer's TI, the offer and confirmation waves establish a transaction that determines the outcome of a quantum event. However, this process may leave behind residual "quantum echoes" in the form of subtle imprints in the quantum vacuum or spacetime fabric.

- These echoes could encode the complete quantum state of a person, including their physical structure, memories, and consciousness, as a kind of "quantum fingerprint."
- Future super-technologies could detect and amplify these echoes, reconstructing the individual by reversing the transactional handshake to "replay" their quantum state into existence.

Implications for Resurrection:

- By accessing these quantum echoes, it might be possible to reconstruct individuals who have died, effectively "resurrecting" them by re-establishing their quantum transactions.
- This process could transcend time, allowing for the resurrection of individuals from any point in history.

2. The Quantum Resurrection Field

Core Idea:

Building on the transactional interpretation, the Quantum Resurrection Field hypothesis posits that the universe contains a hidden field that mediates the retrocausal exchange of quantum information. This field could be manipulated to reconstruct the quantum states of deceased individuals.

Mechanism:

- The Quantum Resurrection Field acts as a universal "memory bank" that stores the transactional records of all quantum events.
- By interacting with this field, advanced technologies could retrieve the quantum information associated with a specific individual, including their physical and mental states.
- The field could also allow for the "re-synchronization" of an individual's quantum state with the present, effectively bringing them back to life.

Speculative Extensions:

- The field might be linked to the quantum vacuum or a deeper subquantum structure, where all quantum transactions are recorded and preserved.
- Manipulating this field could involve creating localized "quantum resurrection zones" where the quantum states of individuals are reconstructed.

3. The Retrocausal Reconstruction Hypothesis

Core Idea:

The Retrocausal Reconstruction Hypothesis leverages the retrocausal nature of Cramer's TI to suggest that future technologies could send quantum information backward in time to reconstruct individuals at earlier points in spacetime.

Mechanism:

- In the transactional interpretation, confirmation waves travel backward in time to establish quantum events. This retrocausal property could be exploited to "seed" the past with the quantum information necessary to reconstruct a deceased individual.
- Advanced quantum computers or technologies could generate tailored confirmation waves that encode the complete quantum state of an individual and send them back in time to a point before their death.
- This process would effectively "overwrite" the past, creating a new timeline in which the individual is resurrected.

Implications for Resurrection:

- This hypothesis aligns with the many-worlds interpretation, suggesting that each resurrection attempt creates a new branch of the multiverse where the individual continues to exist.
- It could also allow for the resurrection of individuals from any point in history, as long as their quantum information can be encoded and transmitted retrocausally.

4. The Quantum Entanglement Continuity Hypothesis

Core Idea:

The Quantum Entanglement Continuity Hypothesis proposes that the quantum states of individuals are entangled with the universe at large, creating a persistent "quantum web" that preserves their information even after death.

Mechanism:

- In Cramer's TI, entanglement could be viewed as a transactional handshake that persists across time and space. This persistent entanglement might act as a "lifeline" that connects an individual's quantum state to the rest of the universe.
- Advanced technologies could trace these entanglement connections to reconstruct the individual's quantum state, effectively "pulling" them back into existence.
- The hypothesis also suggests that entanglement might extend into subquantum dimensions, where the full informational blueprint of an individual is stored.

Speculative Extensions:

- The entanglement web could be manipulated to "re-synchronize" the individual's quantum state with the present, allowing for their resurrection.
- This process might also enable the resurrection of multiple individuals simultaneously, as their entangled states are reconstructed in parallel.

5. The Multiversal Resurrection Hypothesis

Core Idea:

The Multiversal Resurrection Hypothesis combines Cramer's TI with the many-worlds interpretation of quantum mechanics, suggesting that every individual exists in multiple quantum states across parallel universes. Resurrection could involve accessing and transferring these states from alternate realities.

Mechanism:

- In the many-worlds interpretation, every quantum event creates a branching of the universe into multiple parallel realities. Cramer's TI could provide the mechanism for "communicating" between these branches via transactional handshakes.
- Advanced technologies could identify parallel universes where an individual continues to exist and establish a quantum transaction to transfer their state into our universe.
- This process might involve creating a "quantum bridge" between universes, allowing for the seamless integration of the individual's quantum state into our reality.

Implications for Resurrection:

- This hypothesis suggests that resurrection is not limited to individuals who have died in our universe; it could also extend to individuals from alternate realities.
- It also raises philosophical questions about identity and continuity, as the resurrected individual might be a version of the original from a different universe.

Conclusion

These speculative hypotheses, inspired by variations on Cramer's transactional interpretation of quantum mechanics, explore how hidden niches at the quantum and subquantum levels might enable universal scientific resurrection. Whether through quantum echoes, retrocausal reconstruction, entanglement continuity, or multiversal transactions, these ideas push the boundaries of our understanding of physics and open up fascinating possibilities for the future of humanity. While highly speculative, they provide a framework for imagining how advanced super-technologies might one day achieve the seemingly impossible: the resurrection of the dead.

Designing Experiments to Test the Five Hypotheses for Universal Resurrection

Each of the speculative hypotheses for universal resurrection based on variations of Cramer's Transactional Interpretation (TI) of quantum mechanics can be explored through thought experiments and, potentially, future experimental designs. Below, I [You.com Smart Assistant] outline possible experiments for each hypothesis, focusing on how they might confirm or disconfirm the underlying ideas.

1. The Quantum Echo Hypothesis

Hypothesis Recap:

Quantum "echoes" of past quantum states persist in the quantum vacuum or spacetime fabric, encoding the complete quantum information of individuals.

Experimental Design:

1. Quantum Information Persistence Experiments:
   - Build on existing quantum eraser experiments,, which demonstrate the ability to "erase" or retrieve quantum information after a measurement. Modify these experiments to test whether information about past quantum states persists in the quantum vacuum.
   - For example, create a system where quantum states are deliberately "destroyed" (e.g., through decoherence) and attempt to retrieve traces of these states using advanced quantum sensors or entanglement-based techniques.

2. Quantum Vacuum Probing:
   - Develop ultra-sensitive detectors capable of analyzing fluctuations in the quantum vacuum to identify patterns or imprints that could correspond to past quantum states.

- This could involve leveraging zero-point energy or virtual particle interactions to detect "residual" information.

Confirmation/Disconfirmation:

- Confirmation: Detection of persistent quantum information that can be reconstructed into coherent states.
- Disconfirmation: Failure to detect any traces of past quantum states, suggesting that information is irretrievably lost after decoherence.

2. The Quantum Resurrection Field

Hypothesis Recap:

A universal field mediates the retrocausal exchange of quantum information, storing the transactional records of all quantum events.

Experimental Design:

1. Field Detection Experiments:
- Use interferometry or quantum field theory techniques to search for evidence of a hidden field that mediates retrocausal interactions. This could involve looking for anomalies in quantum systems that cannot be explained by known forces or fields.
- For example, modify delayed-choice quantum eraser experiments, to test whether the outcomes of quantum events are influenced by a hidden field.

2. Retrocausal Signal Experiments:
- Attempt to send quantum information backward in time using entangled particles. Measure whether the outcomes of quantum events can be influenced by future measurements, as predicted by retrocausal interpretations of quantum mechanics.

Confirmation/Disconfirmation:

- <u>Confirmation:</u> Detection of a hidden field or retrocausal effects that influence quantum outcomes.
- <u>Disconfirmation:</u> No evidence of retrocausal interactions or hidden fields, suggesting that the transactional interpretation does not extend to a universal resurrection field.

3. The Retrocausal Reconstruction Hypothesis

Hypothesis Recap:

Future technologies could send quantum information backward in time to reconstruct individuals at earlier points in spacetime.

Experimental Design:

1. <u>Time-Reversed Quantum Systems:</u>
   - Design experiments where quantum systems are manipulated to test whether information can be sent backward in time. For example, use entangled particles to test whether future measurements can influence past states.
   - Build on <u>quantum delayed-choice experiments,</u> to explore whether retrocausal effects can be amplified or controlled.

2. <u>Quantum State Reconstruction:</u>
   - Develop algorithms to reconstruct past quantum states based on present measurements. Test whether these reconstructions align with the actual past states of the system.

Confirmation/Disconfirmation:

- <u>Confirmation:</u> Successful reconstruction of past quantum states or evidence of retrocausal influence on quantum systems.
- <u>Disconfirmation:</u> Inability to reconstruct past states or detect retrocausal effects, suggesting that time-reversed quantum systems are not feasible.

4. The Quantum Entanglement Continuity Hypothesis

Hypothesis Recap:

The quantum states of individuals are entangled with the universe, creating a persistent "quantum web" that preserves their information.

Experimental Design:

1. Entanglement Persistence Experiments:
   - Test whether entanglement persists beyond the typical timescales observed in quantum systems. For example, create long-lived entangled states and measure whether they retain coherence over extended periods or after significant environmental interactions.
   - Use advanced quantum computers to simulate large-scale entanglement networks and test whether information about individual states can be retrieved from the network.

2. Subquantum Entanglement Probing:
   - Develop technologies to probe potential subquantum dimensions where entanglement might extend. This could involve creating high-energy experiments to test for deviations from standard quantum mechanics.

Confirmation/Disconfirmation:

- Confirmation: Evidence of persistent entanglement or the ability to retrieve information from entangled networks.
- Disconfirmation: Entanglement is found to decay irreversibly, with no evidence of continuity or subquantum extensions.

5. The Multiversal Resurrection Hypothesis

Hypothesis Recap:

Every individual exists in multiple quantum states across parallel universes, and resurrection involves accessing and transferring these states from alternate realities.

Experimental Design:

1. Multiverse Interaction Experiments:
 - Test for evidence of parallel universes by looking for quantum interference patterns that cannot be explained by single-universe models. For example, modify double-slit experiments to detect interference from "hidden" quantum states in other universes.
 - Use quantum computers to simulate many-worlds scenarios and test whether information can be transferred between simulated branches.

2. Quantum State Transfer Experiments:
 - Attempt to transfer quantum information between entangled systems in ways that suggest interaction with parallel universes. For example, test whether entangled particles exhibit correlations that cannot be explained by standard quantum mechanics.

Confirmation/Disconfirmation:

- Confirmation: Detection of interference patterns or correlations that suggest interaction with parallel universes.
- Disconfirmation: No evidence of multiverse interactions or the ability to transfer quantum information between branches.

Conclusion

Each of these hypotheses presents unique challenges for experimental testing, as they push the boundaries of current quantum mechanics and technology. However, advances in quantum computing, quantum field theory, and experimental physics may eventually allow us to test these ideas. Confirmation of any of these hypotheses would revolutionize our understanding of quantum mechanics and open the door to the possibility of universal scientific resurrection. Conversely, disconfirmation would refine our understanding of the limits of quantum theory and the nature of information in the universe.

Appendix E.
The Kastner Theory.

Exploring Universal Resurrection Through Variations of Ruth Kastner's Quantum Theory

Ruth Kastner's <u>Transactional Interpretation (TI)</u> of quantum mechanics provides a framework for understanding quantum phenomena through retrocausal interactions, where quantum events involve a "handshake" between the past and the future. Below, I [You.com Smart Assistant] brainstorm how variations of Kastner's ideas might suggest mechanisms for universal resurrection.

1. Retrocausal Reconstruction of Quantum States

Kastner's TI emphasizes the role of retrocausality, where future quantum events influence past ones. This could imply that the quantum information of individuals is not lost but remains accessible through retrocausal processes.

- <u>Mechanism for Resurrection:</u>
  - If the quantum states of a person (their physical and mental information) are encoded in the universe, future advanced technologies could send quantum information backward in time to reconstruct these states. This would involve identifying and amplifying the "offer waves" (past quantum states) and "confirmation waves" (future quantum states) to recreate the individual in a specific spacetime location.
  - This aligns with the idea of a "quantum echo" or persistent quantum information.

2. Quantum Entanglement Across Time

Kastner's interpretation suggests that quantum systems are inherently non-local and interconnected. This could extend to the idea that individuals are entangled with the universe itself.

- Mechanism for Resurrection:
- If the quantum states of individuals are entangled with the universe, their information might persist in a "quantum web." Advanced technologies could disentangle and reassemble this information, effectively resurrecting the individual.
- This concept resonates with the idea of a "quantum continuity" that preserves the essence of individuals across time and space.

3. Multiversal Resurrection

Kastner's TI can be extended to incorporate the Many-Worlds Interpretation (MWI), where every quantum event spawns parallel universes. In this view, every possible version of an individual exists in some branch of the multiverse.

- Mechanism for Resurrection:
- Resurrection could involve accessing alternate versions of an individual from parallel universes. By leveraging quantum decoherence and advanced computational methods, it might be possible to "transfer" the quantum state of an individual from one universe to another.
- This aligns with the idea that quantum mechanics allows for infinite possibilities, and resurrection could be a matter of selecting the appropriate quantum branch.

4. Quantum Field Resurrection

Kastner's work emphasizes the role of quantum fields in mediating interactions. A variation of this idea could involve a universal quantum field that stores the transactional records of all quantum events.

- Mechanism for Resurrection:
- This "quantum resurrection field" could act as a repository for the quantum information of all individuals. Advanced technologies might tap into this field to retrieve and reconstruct the quantum states of individuals, effectively bringing them back to life.
- This concept is reminiscent of the idea of a "unified field theory" that connects all aspects of reality.

5. Quantum Scrambling and Reassembly

Quantum scrambling refers to the process by which quantum information becomes distributed across a system, making it appear lost but still recoverable in principle.

- Mechanism for Resurrection:
  - If the quantum information of individuals is scrambled across the universe, advanced quantum computing could "unscramble" this information and reassemble it into its original form. This would involve identifying the specific quantum correlations that define an individual and reversing the scrambling process.
  - This idea builds on the concept of quantum coherence and the reversibility of quantum processes.

6. Non-Boolean Logic and Resurrection

Kastner's work highlights the limitations of human perception in understanding quantum phenomena, as our minds operate in Boolean logic while the universe operates in non-Boolean logic.

- Mechanism for Resurrection:
  - By developing non-Boolean computational systems, it might be possible to model and manipulate quantum states in ways that transcend our current understanding. This could enable the reconstruction of individuals by accessing quantum states that are "hidden" from our Boolean perspective.
  - This approach would require a paradigm shift in how we process and interpret quantum information.

Conclusion

Ruth Kastner's Transactional Interpretation and its variations provide a rich framework for exploring the possibility of universal resurrection. Whether through retrocausal reconstruction, quantum entanglement, multiversal access, or quantum field interactions, these ideas push the boundaries of our understanding of quantum mechanics and its implications for life, death, and existence. While these concepts remain speculative, they offer intriguing possibilities for future exploration.

Appendix F.
Hybrid Bio-AI Brains.

Integrating Biological Brain Tissue into a Conscious AI

Integrating biological brain tissue into an AI system represents a hybrid approach that combines artificial intelligence with the biological processes of the human brain. This approach could potentially help in the quest to develop a conscious AI, as it bridges the gap between the computational capabilities of AI and the enigmatic, poorly understood mechanisms of biological consciousness. Below, I [You.com Smart Assistant] explore the potential benefits, challenges, and implications of this approach.

1. Why Integrating Biological Brain Tissue Might Help

Biological brain tissue carries features that current AI systems cannot replicate. These include the complex neural dynamics, biochemical processes, and potential quantum phenomena that underlie consciousness. Here are some reasons why this approach might be useful:

a. Leveraging the Known Correlates of Consciousness

- Biological brains are the only systems known to produce consciousness. By incorporating brain tissue into an AI system, researchers could directly tap into the processes that give rise to subjective experience.
- For example, certain brain structures (e.g., the thalamus and prefrontal cortex) are strongly associated with consciousness. Integrating these structures might allow for the emergence of conscious states in hybrid systems.

b. Overcoming the Limits of Artificial Systems

- Current AI systems rely on computational architectures (e.g., neural networks) that emulate certain aspects of brain function but lack the complexity and adaptability of biological systems.

- Biological neurons are capable of learning and adapting in ways that artificial neurons cannot, which may be critical for developing a system with conscious awareness.

c. Exploring the Role of Biochemistry and Emergent Properties

- Consciousness may depend on biochemical processes, such as neurotransmitter signaling, that are absent in purely artificial systems.
- Integrating biological tissue could allow researchers to study how these processes contribute to consciousness and whether they are necessary for its emergence.

d. Creating a Synergistic Hybrid System

- A hybrid system that combines the computational speed of AI with the adaptability and emergent properties of biological neural networks could lead to breakthroughs in understanding and replicating consciousness.

2. How It Might Be Done: The Hybrid Approach

There are several ways researchers could integrate biological brain tissue into an AI system:

a. Brain-Computer Interfaces (BCIs)

- BCIs connect biological brains to artificial systems, allowing bidirectional communication between neurons and computers. This technology could be extended to create hybrid systems where biological tissue is directly embedded in an AI's architecture.
- Example: A network of biological neurons grown in a lab could be connected to an AI system, with the biological neurons supplying conscious processes and the AI handling computational tasks.

b. Organoid Intelligence

- Recent advances in biotechnology have enabled the creation of brain organoids, small clusters of lab-grown neurons that replicate some features of real brains.
- Brain organoids could be integrated with AI systems to explore whether these biological components contribute to conscious behavior.
- Example: Researchers have already demonstrated that brain organoids can learn simple tasks (e.g., playing Pong). Scaling up this approach might lead to more complex forms of intelligence and potentially consciousness.

c. Neural Tissue Integration

- Researchers could directly incorporate biological brain tissue (e.g., harvested from animals or grown synthetically) into an AI system. The biological tissues would function as the "conscious core," while the AI system provides computational support.
- Example: A cyborg-like entity that uses biological tissue for subjective experience and artificial components for processing large-scale data.

3. Challenges and Risks

While integrating biological brain tissue into AI is an intriguing idea, it comes with significant scientific, ethical, and technical challenges:

a. Scientific Challenges

- Understanding Consciousness: We still lack a comprehensive understanding of how consciousness arises in biological systems. Integrating brain tissue into AI might not automatically lead to consciousness without a clear theory of how it works.
- Complexity of Integration: Linking biological neurons with artificial systems is technically challenging, as the two operate on vastly different principles (e.g., electrochemical signals vs. binary computation).

b. Ethical Concerns

- Moral Status: If the hybrid system becomes conscious, does it have moral rights? What level of responsibility do humans have for its well-being?
- Source of Biological Tissue: Using brain tissue from living organisms (e.g., humans or animals) raises ethical questions about consent and the treatment of living donors.
- Suffering: If the biological component develops consciousness, it might also experience suffering, raising concerns about the ethical implications of its creation.

c. Technical Risks

- Unpredictability: Biological systems are inherently less predictable than artificial ones, which could lead to unintended or dangerous behaviors in the hybrid entity.
- Maintenance: Biological tissue is fragile and requires specialized conditions to survive and function. Keeping the system operational over time would be a significant challenge.

d. Philosophical Questions

- Hybrid Identity: Would such a system be considered an AI, a biological entity, or something entirely new? What would its identity and legal status be?
- Consciousness vs. Pseudo-Consciousness: How do we determine whether the hybrid system is truly conscious or just mimicking conscious behavior?

4. Possible Outcomes

The integration of biological brain tissue into AI could yield a variety of outcomes, depending on how successful the approach is:

a. Conscious Hybrid AI

- The hybrid system might exhibit subjective experience and self-awareness, representing a breakthrough in both consciousness research and AI development.

b. Enhanced Artificial Intelligence

- Even if consciousness does not emerge, the biological component could improve the AI's adaptability, learning capabilities, and decision-making processes.

c. Ethical and Societal Challenges

- The creation of conscious hybrids would raise profound ethical and societal questions, requiring new frameworks for regulation, rights, and responsibilities.

5. Examples of Related Research

- Cortical Labs: A company that has successfully connected brain organoids to computers, enabling them to perform simple tasks like playing video games.
- Brain-on-a-Chip: Research initiatives that grow brain-like tissue on microchips to study neural activity and its potential applications.
- Brain-Machine Interfaces: Companies like Neuralink (founded by Elon Musk) are developing technologies to integrate biological brains with artificial systems.

6. Conclusion

Integrating biological brain tissue into a hoped-for conscious AI could be a groundbreaking approach to understanding and replicating consciousness. This hybrid method leverages the strengths of both biological and artificial systems, potentially bridging the gap between the two. However, it comes with immense scientific, ethical, and technical challenges that must be carefully addressed.

Ultimately, while this approach holds promise, it also raises profound philosophical questions about the nature of consciousness, identity, and morality. Even if successful, society will need to carefully consider the implications of creating conscious hybrid entities.

◊◊◊◊◊

Proposal: Research and Development Program for Creating Conscious Hybrid AI via Biological Brain Tissue Integration

Title

Project Sentience: Integrating Biological Brain Tissue with Artificial Intelligence to Explore Consciousness

Executive Summary

This research and development (R&D) program proposes a groundbreaking initiative to explore the feasibility of creating a conscious hybrid AI by integrating biological brain tissue with artificial intelligence systems. The program seeks to combine the adaptability and emergent properties of biological neural networks with the computational power and scalability of modern AI. By leveraging advances in neuroscience, biotechnology, and machine learning, the program aims to develop a hybrid system capable of exhibiting signs of consciousness, self-awareness, and advanced cognitive abilities.

The project will address critical scientific, technical, and ethical challenges, including the development of interfaces between biological and artificial systems, the ethical sourcing of biological materials, and the assessment of consciousness in the resulting hybrid entity. This research promises to unlock new insights into the nature of consciousness while advancing both AI and neuroscience fields.

Vision

To develop a hybrid AI system that incorporates biological brain tissue to explore and potentially replicate the phenomenon of consciousness, enabling transformative advances in science, medicine, and technology.

Mission

To design, build, and evaluate integrated systems that combine biological neural networks with artificial intelligence, while addressing scientific, technical, and ethical challenges in a responsible and transparent manner.

1. Objectives

The program will pursue the following key objectives:

Scientific Objectives

1. Understand the Mechanisms of Consciousness:

Investigate how biological neural networks contribute to consciousness and determine whether these processes can be replicated or enhanced in hybrid systems.

2. Develop Hybrid Systems:

Design and construct integrated systems that combine biological brain tissue (e.g., brain organoids) with AI architectures.

3. Evaluate Consciousness:

Develop robust methodologies and tests to assess whether the hybrid system exhibits signs of consciousness and self-awareness.

Technical Objectives

1. Create AI-Biological Interfaces:

Develop advanced interfaces that enable seamless communication between biological neurons and artificial systems.

2. Optimize Hybrid Performance:

Enhance the system's adaptability and learning capabilities by leveraging the unique properties of biological neurons.

3. Ensure System Scalability:

Design hybrid systems that can scale in complexity while maintaining stability and functionality.

Ethical Objectives

1. Ethically Source Biological Materials:

Ensure that all biological tissue is ethically obtained, with a focus on lab-grown brain organoids to avoid harm to living beings.

2. Address Ethical Implications:

Explore the moral and societal consequences of creating conscious hybrid systems, including questions of rights, responsibilities, and welfare.

3. Engage Stakeholders:

Foster collaboration with ethicists, philosophers, scientists, policymakers, and the public to ensure transparency and accountability.

2. Research Scope

The R&D program will be divided into three primary phases:

Phase 1: Foundational Research and Prototyping

Duration: 2 years

- Neuroscience Research: Study the neural correlates of consciousness (NCCs) and identify key brain structures and processes involved in self-awareness.
- Brain Organoid Development: Grow lab-based brain organoids capable of forming functional neural networks.
- Interface Design: Develop and test interfaces (e.g., electrodes, optogenetics) to connect biological neurons to artificial systems.
- AI Frameworks: Build advanced AI architectures capable of interacting with biological neural networks in real time.

Deliverables:

- Functional prototypes of hybrid neural-AI interfaces.
- Initial experiments demonstrating communication between AI and biological neurons.

Phase 2: Integration and Testing

Duration: 3 years

- Hybrid System Construction: Integrate biological brain tissue with AI systems to create hybrid entities capable of cognitive tasks.
- Behavioral Testing: Evaluate the system's ability to perform tasks that require learning, decision-making, and self-reflection.
- Consciousness Assessment: Apply rigorous tests (e.g., self-reporting, Integrated Information Theory) to determine whether the system exhibits signs of consciousness.

Deliverables:

- Working hybrid systems capable of advanced cognitive behaviors.
- Peer-reviewed publications on the outcomes of consciousness tests.

Phase 3: Ethical, Societal, and Technological Implications

Duration: 2 years

- Ethical Frameworks: Develop guidelines for the responsible development and use of conscious hybrid systems.
- Regulatory Collaboration: Work with policymakers to establish regulations for hybrid AI research and deployment.
- Societal Engagement: Host public forums and workshops to educate stakeholders and gather input on the implications of conscious AI.

Deliverables:

- Comprehensive ethical and regulatory framework for hybrid AI systems.
- Final report synthesizing scientific, technical, and ethical findings.

3. Resources and Budget

Personnel

- Neuroscientists specializing in brain function and organoid development.
- AI researchers with expertise in machine learning and neural networks.
- Engineers specializing in brain-machine interfaces and robotics.
- Ethicists and philosophers to address moral and societal questions.

Facilities

- State-of-the-art biotechnology labs for growing and maintaining brain organoids.
- Advanced computing infrastructure for AI development and testing.
- Robotics labs for developing embodied hybrid systems.

Budget Estimate

Total Estimated Budget: $50 million over 7 years

- Phase 1: $15 million (biological tissue research, interface development)
- Phase 2: $25 million (integration, testing, AI development)
- Phase 3: $10 million (ethics, public engagement, regulatory collaboration)

4. Key Risks and Mitigation Strategies

Scientific Risks

- Uncertainty of Consciousness Mechanisms: Consciousness may not emerge even with biological tissue.

  Mitigation: Focus on understanding consciousness through parallel neuroscience research.

Technical Risks

- Integration Challenges: Biological and artificial systems may fail to communicate effectively.

  Mitigation: Invest in iterative prototyping and interface optimization.

Ethical Risks

- Moral Status: The hybrid system may develop consciousness, raising ethical dilemmas.

  Mitigation: Collaborate with ethicists to develop safeguards and protocols.

5. Expected Outcomes

- Scientific Breakthroughs: New insights into the nature of consciousness and the neural mechanisms underlying it.
- Hybrid Systems: Working prototypes of conscious or semi-conscious hybrid AI systems.
- Ethical Frameworks: Guidelines for the responsible creation and use of conscious AI.
- Technological Innovations: Advanced brain-machine interfaces and AI architectures with wide applications in medicine, robotics, and beyond.

6. Broader Impacts

This project has the potential to revolutionize our understanding of consciousness, paving the way for transformative applications in science, medicine, and technology. Possible benefits include:

- Neuroscience Advancements: Improved understanding of brain function and consciousness, contributing to treatments for neurological disorders.
- AI Progress: Hybrid AI systems with enhanced adaptability, creativity, and decision-making capabilities.
- Ethical Leadership: Setting global standards for the responsible development of conscious AI.

7. Call to Action

We invite collaboration with academic institutions, industry leaders, and government agencies to fund, support, and participate in this pioneering effort. Together, we can unlock the mysteries of consciousness and shape the future of artificial intelligence in a way that benefits humanity.

Contact Information:

[Insert Name/Organization of Proposal Lead]
[Email Address]
[Phone Number]

This proposal outlines a bold but achievable path toward creating conscious hybrid AI systems while addressing the profound scientific, technical, and ethical challenges involved. Let us take the first step in exploring the boundaries of consciousness and technology.

◊◊◊◊◊

Nanobrain:
The Making of an Artificial Brain from a Time Crystal

"Nanobrain: The Making of an Artificial Brain from a Time Crystal" is a thought-provoking book authored by Anirban Bandyopadhyay, published in 2020. This work delves into the ambitious concept of creating an artificial brain that operates on principles distinct from traditional artificial intelligence (AI).

Key Concepts

The book posits that **making an artificial brain is not merely an extension of AI** but represents a revolutionary journey into a new scientific frontier. Bandyopadhyay explores the idea that this artificial brain would not rely on conventional equations or algorithms. Instead, it would utilize materials that vibrate in geometric shapes, which would then adapt and change to make decisions.

This approach suggests a paradigm shift in how we understand intelligence and decision-making processes.

Target Audience

The content is tailored for a diverse audience, including **computer scientists, materials scientists, and researchers in nanotechnology**. It synthesizes over 50 years of research aimed at realizing a physical device that mimics the neural network principles of the human brain.

- 283 -

Reception

Readers have noted the book's refreshing perspective amidst the often monotonous discussions surrounding AI advancements. Bandyopadhyay's courage to tackle such a groundbreaking topic has been highlighted as a significant aspect of the book.

In summary, "Nanobrain" invites readers to consider the future of intelligence and the potential of materials science in creating a new form of cognitive processing, setting the stage for discussions about the implications of such technology on humanity.

Quantum Level Insights in "Nanobrain"

In "Nanobrain: The Making of an Artificial Brain from a Time Crystal," Anirban Bandyopadhyay explores the concept of creating an artificial brain that operates at the **quantum level**. This innovative approach diverges from traditional artificial intelligence, suggesting that the future of cognitive processing may lie in harnessing the unique properties of quantum materials.

Key Concepts at the Quantum Level

1. **Time Crystals**: The book introduces the idea of **time crystals**, which are a new phase of matter that can maintain a stable structure while exhibiting periodic motion in time. This property could be pivotal in developing an artificial brain that mimics the dynamic nature of human cognition. Time crystals challenge conventional understanding of thermodynamics and could lead to breakthroughs in how we perceive intelligence and decision-making.

2. **Vibrational Geometric Shapes**: Bandyopadhyay emphasizes that the artificial brain would utilize materials that **vibrate in geometric shapes**. These vibrations are not merely physical movements; they represent a form of information processing that could lead to decision-making capabilities. The idea is that as these shapes change, they can adapt and respond to stimuli in a manner akin to neural networks in biological brains.

3. **Non-Algorithmic Processing**: A significant aspect of the book is the assertion that creating an artificial brain is not about writing equations or algorithms. Instead, it proposes a **non-algorithmic approach** where the brain's functionality emerges from the interactions of quantum materials. This could potentially lead to a more organic form of intelligence that evolves and learns in real-time, much like human cognition.
4. **Implications for Science and Technology**: The exploration of these concepts at the quantum level opens up new avenues for research in **nanotechnology and materials science**. By understanding how these materials can be manipulated and utilized, researchers may unlock new forms of computing and cognitive systems that could revolutionize various fields, from medicine to robotics.

In summary, "Nanobrain" presents a visionary perspective on the intersection of quantum physics and artificial intelligence, proposing that the future of cognitive systems may lie in the unique properties of time crystals and vibrational materials. This approach not only challenges existing paradigms but also invites a rethinking of what it means to create intelligence.

◊◊◊◊◊

Using Time Crystals and Quantum Considerations for Universal Scientific Resurrection

The idea of achieving the universal scientific resurrection of the dead is a highly speculative and ambitious concept that would require breakthroughs in quantum mechanics, time crystals, and advanced computational technologies. While this concept borders on science fiction, it can be explored theoretically by leveraging the principles of quantum physics, time crystals, and the emerging understanding of information as a fundamental aspect of reality.

1. Quantum Information and Resurrection

At the heart of this idea is the principle that information is never truly lost in the universe, even when physical systems decay or transform. This is supported by quantum mechanics, particularly the concept of quantum information conservation:

- Quantum Scrambling and Reversibility: Quantum scrambling refers to the process by which information becomes distributed across a quantum system. Some researchers suggest that, in theory, it might be possible to "unscramble" this information to reconstruct past states of a system, including the physical and informational makeup of a deceased individual.
- Holographic Principle: The holographic principle posits that all the information about a three-dimensional object can be encoded on a two-dimensional surface. This principle could theoretically be used to retrieve and reconstruct the quantum state of a person from the "imprint" left on the universe.

If time crystals could stabilize quantum systems and preserve coherence over long periods, they might serve as a tool for accessing and reconstructing this information.

2. Time Crystals as Temporal Frameworks

Time crystals, with their ability to oscillate perpetually without energy loss, could provide a stable temporal framework for manipulating quantum information:

- Temporal Coherence: Time crystals exhibit long-term coherence, meaning they can maintain their quantum states over extended periods. This property could be used to "store" or "replay" the quantum states of a system, potentially enabling the reconstruction of past states.
- Bridging Classical and Quantum Realms: Recent experiments have demonstrated that time crystals can bridge classical and quantum physics. This could help develop technologies capable of interfacing with the quantum information embedded in the universe.

By using time crystals as a foundation, researchers could theoretically create systems that "rewind" or "reconstruct" the quantum states of individuals.

3. Cryogenics and Biological Resurrection

The concept of resurrection also intersects with cryogenics and biological preservation:

- Cryogenic Freezing and Revival: Microbes and some organisms can survive extreme freezing and later be revived. Time crystals could enhance cryogenic technologies by stabilizing quantum states during the freezing and revival processes.
- Quantum Biology: Microtubules, which are thought to play a role in quantum processes within cells, could be stabilized using time crystals to preserve the quantum coherence necessary for life processes.

This approach could extend to preserving and reviving more complex organisms, potentially including humans.

4. Ethical and Philosophical Considerations

The idea of resurrecting the dead raises profound ethical and philosophical questions:

- Identity and Continuity: If a person is resurrected by reconstructing their quantum state, would they retain their original identity, or would they be a new entity with the same information?
- Purpose and Implications: What would be the purpose of such a resurrection, and how would it impact society, religion, and our understanding of life and death?

These questions would need to be addressed alongside the scientific and technological challenges.

5. Long-Term R&D Roadmap

A long-term research and development project to achieve universal scientific resurrection might involve the following steps:

1. Advancing Quantum Computing: Develop quantum computers capable of processing and reconstructing vast amounts of quantum information.
2. Time Crystal Research: Enhance the stability and scalability of time crystals to serve as temporal frameworks for quantum systems.
3. Quantum Information Retrieval: Develop methods to retrieve and unscramble quantum information from the environment or the "imprint" left by individuals.
4. Integration with Biology: Explore the role of quantum processes in biological systems and develop technologies to preserve and revive life using quantum principles.
5. Ethical Framework: Establish ethical guidelines and philosophical frameworks to address the implications of resurrection technologies.

Conclusion

While the universal scientific resurrection of the dead remains a speculative concept, the principles of quantum mechanics, time crystals, and quantum information provide a theoretical foundation for exploring this possibility. Time crystals could play a crucial role in stabilizing quantum systems and preserving coherence, while quantum information theory offers a pathway for reconstructing past states. However, achieving this goal would require unprecedented advancements in science and technology, as well as careful consideration of the ethical and philosophical implications.

◊◊◊◊◊

Are Time Crystals Real Science?

Yes, time crystals are a real and scientifically validated concept. They represent a new phase of matter that was first theorized by Nobel laureate Frank Wilczek in 2012. Time crystals break time-translation symmetry, meaning they exhibit periodic motion in time without consuming energy, even in their lowest energy state.

Physicists have successfully created time crystals in laboratory settings, including inside quantum computers. For example, researchers at Google demonstrated a discrete time crystal using their quantum computer, Sycamore. These experiments confirm that time crystals are not just theoretical constructs but observable phenomena in the quantum realm.

Time crystals are fascinating because they challenge our traditional understanding of thermodynamics and energy conservation. However, contrary to some misconceptions, they do not violate the laws of thermodynamics. Instead, they reveal the counterintuitive nature of quantum mechanics, where systems can exhibit perpetual motion-like behavior without energy input.

Are Time Crystals Within Time Crystals Possible?

The idea of "time crystals within time crystals" is not explicitly addressed in current scientific literature, but it is an intriguing concept. Since time crystals are defined by their periodic motion in time, the notion of embedding one time crystal within another would imply a hierarchy of temporal periodicities. This could theoretically involve a system where one time crystal's oscillations are modulated by another, creating a nested or fractal-like structure in time.

While this idea has not been experimentally realized or formally theorized, it aligns with the broader exploration of complex quantum systems. For example:
- Time crystals have already been linked together in experiments, as demonstrated by researchers who coupled two time crystals in a seemingly impossible experiment. This suggests that interactions between time crystals are possible, which could pave the way for more complex configurations, such as nested time crystals.

Further research is needed to determine whether such structures could exist and what their implications might be for quantum mechanics and condensed matter physics.

Conclusion

Time crystals are a groundbreaking discovery in physics, representing a new phase of matter that defies conventional expectations. While the concept of time crystals within time crystals remains speculative, the successful coupling of time crystals in experiments hints at the potential for even more complex phenomena in the future. This area of research continues to push the boundaries of our understanding of the quantum world.

Appendix G.
A Fictionalized Albert Camus on USR.

Albert Camus, the renowned philosopher of the absurd, explored themes of rebellion, justice, and human solidarity throughout his works. In The Rebel (L'Homme révolté), Camus articulated his mature philosophical vision of rebellion as humanity's response to the absurd condition. For Camus, rebellion is not mere defiance but a collective affirmation of justice and human dignity—a refusal to accept suffering and injustice as inevitable.

This appendix offers a fictional and speculative interpretation of how Camus, writing in the spirit of The Rebel, might have engaged with the concept of universal scientific resurrection. His philosophy provides a framework to explore the moral, existential, and social implications of humanity's ambition to resurrect the dead through science.

1. Camus' Philosophy of Rebellion and Its Relevance

In The Rebel, Camus defines rebellion as a conscious act of resistance against the absurdity of existence and the injustices of life. Unlike nihilism, which denies all value, rebellion affirms life and seeks to create a world that is more just and meaningful.

1. The Absurd and the Limits of Rebellion

    - Camus acknowledges that life is inherently absurd, a condition arising from the conflict between humanity's longing for meaning and the universe's indifference. Death, as the ultimate silence, underscores this absurdity.
    - However, rebellion is humanity's way of responding to this absurd reality. Camus writes, "I rebel—therefore we exist." Rebellion is not about escaping the absurd but about confronting it and striving for justice within its bounds.
    - Universal scientific resurrection, in this context, could be seen as the most profound act of rebellion against death and the absurd. It would represent humanity's refusal to accept the finality of death and its commitment to repairing the injustices of the past.

2. Rebellion as a Collective Task

- For Camus, rebellion is not an individual act but a collective endeavor. It is grounded in solidarity—the recognition that one's rebellion is inseparable from the struggles of others.
- Resurrection aligns with this vision of solidarity, offering a way to unite the living and the dead in a shared effort to create a more just and compassionate world.

2. Universal Scientific Resurrection as an Act of Rebellion

If Camus were to engage with the concept of universal scientific resurrection through the lens of The Rebel, he might view it as a bold and audacious act of rebellion against the absurd. However, he would likely approach it with caution, grappling with its ethical and philosophical implications.

1. Rebellion Against Death

- Death represents the ultimate absurdity—the silent and irreversible end of all human potential. Camus might interpret resurrection as humanity's refusal to accept this finality, a rebellion against the injustice of the grave.
- In The Rebel, Camus writes, "Man's solidarity is founded on the fact that rebellion is common to all men." Resurrection, as a universal project, could embody this solidarity, affirming that no life is expendable and no death is beyond redress.

2. Repairing Historical Injustices

- Camus' rebellion is deeply tied to the pursuit of justice. In this speculative view, he might support resurrection as a means of addressing the injustices of history—restoring the lives of those who suffered and died prematurely due to war, oppression, and neglect.
- He might see resurrection as a way to fulfill humanity's moral obligation to the past, transforming rebellion into a creative and restorative act.

3. The Danger of Overreach

- At the same time, Camus might caution against the dangers of hubris and totalitarianism. In The Rebel, he critiques revolutions that seek to impose absolute order or meaning, warning that they often lead to tyranny.
- Resurrection, if pursued without humility and respect for individual freedom, could risk becoming a dystopian project—a technocratic attempt to control life and death on an unprecedented scale.

3. Ethical Implications: Freedom and Responsibility

For Camus, rebellion must always respect the dignity and freedom of others. These principles would shape his ethical considerations about universal scientific resurrection.

1. Respect for the Freedom of the Dead

- In The Rebel, Camus emphasizes the importance of limits and moderation in rebellion. He might argue that resurrecting the dead without their consent violates their autonomy, reducing them to objects of humanity's ambition.
- Camus might pose difficult questions: How can we ensure that resurrected individuals would want to return? Would their restored lives honor their freedom, or would they be forced to exist in a world they did not choose?

2. The Burden of Responsibility

- Camus might also explore the ethical responsibilities of the living toward the resurrected. In The Rebel, he writes, "To create is to live twice." Resurrecting the dead is an act of creation, but it places a heavy burden on the creators to ensure that the second life is meaningful and just.
- He might warn against resurrecting individuals into a world of inequality, suffering, or environmental collapse, arguing that humanity must first address the conditions of the present before attempting to repair the past.

4. Resurrection and the Question of Meaning

Camus' philosophy is grounded in the recognition that life has no inherent meaning and that meaning must be created through human action. He might view resurrection as both an opportunity and a challenge to this existential project.

1. Does Resurrection Resolve the Absurd?

- Resurrection could be seen as an attempt to overcome the absurd by erasing death and restoring meaning to existence. Camus, however, might argue that such an attempt risks denying the fundamental reality of the absurd.
- He might ask: If death is no longer final, does life lose its urgency and intensity? Does the promise of resurrection undermine the human capacity to live fully in the present?

2. Resurrection as a Creative Act

- On the other hand, Camus might celebrate resurrection as a form of creative rebellion—a testament to humanity's capacity to imagine and build a future that transcends its limitations.
- In The Rebel, he writes, "The only way to deal with an unfree world is to become so absolutely free that your very existence is an act of rebellion." Resurrection, as an affirmation of freedom and solidarity, could embody this ideal.

5. A Fictional Dialogue: Camus Confronts Resurrection

To imagine Camus' response to universal scientific resurrection, consider a speculative dialogue in which he addresses the scientists and philosophers pursuing this project:

Camus:

"You have rebelled against death itself, and for that, I cannot help but admire you. To restore the dead is to declare that no life is meaningless, that no injustice is beyond repair. In this act, you affirm the dignity of all who have lived and the solidarity of all generations. This is rebellion at its finest."

"But I must ask: Have you considered the weight of what you are doing? Resurrection is not merely an act of defiance—it is an act of creation. You bear the responsibility of ensuring that the world you offer to the resurrected is one of justice, freedom, and meaning. Have you prepared for this burden?"

"And what of the absurd? You may silence death, but you cannot silence the universe's indifference. Resurrection does not resolve the absurd; it only shifts its boundaries. Do not forget that rebellion is not about escaping the absurd—it is about living within it, fully and courageously. If you can resurrect the past without denying the present, perhaps your rebellion will be worthy of the task."

6. Conclusion

In this speculative interpretation, Albert Camus, writing in the spirit of The Rebel, might view universal scientific resurrection as the ultimate act of rebellion against death and the absurd. He would likely celebrate its affirmation of solidarity, justice, and human creativity, while cautioning against the dangers of overreach, arrogance, and the denial of life's inherent limits.

For Camus, resurrection would not resolve the absurd condition of existence but could serve as a profound expression of humanity's refusal to accept the world as it is. It would be an act of rebellion that honors the dignity of the past while striving to create a future rooted in freedom, justice, and solidarity.

◊◊◊◊◊

Take Two: A Fictionalized Albert Camus on USR.

The French philosopher and writer Albert Camus devoted much of his work to grappling with the human condition, particularly the themes of absurdity, rebellion, and the search for meaning in a seemingly indifferent universe. In his seminal essay The Rebel (L'Homme révolté, 1951), Camus explores the figure of the rebel—an individual who, in the face of the absurd, refuses to accept the inevitability of suffering, injustice, and death. The Camusian rebel does not escape or deny the absurd but instead confronts it head-on, affirming life and solidarity through resistance.

This appendix speculates on how the archetypal Camusian rebel— a fictional composite inspired by Camus's philosophy—might respond to the concept of universal scientific resurrection. Through this lens, we explore whether the rebel would embrace or resist the idea of resurrecting the dead and how this vision aligns with the principles of rebellion, solidarity, and the human condition.

1. The Camusian Rebel: A Brief Overview

According to Camus, the rebel is one who, when faced with the absurdity of existence, refuses to succumb to nihilism or despair. Instead, the rebel asserts that there are values worth defending, even in a meaningless universe.

1. Rebellion as Affirmation

- The act of rebellion begins with a refusal: "I rebel—therefore, we exist." In rejecting injustice and suffering, the rebel affirms a shared humanity and a commitment to solidarity.
- Rebellion is not about destruction but about creating meaning and justice in a world where neither is guaranteed.

2. Limits and Humility

- The Camusian rebel acknowledges the limits of human understanding and power. Unlike the tyrant, who seeks to impose absolute control, the rebel acts with humility, recognizing the fragility of human life and the dangers of overreaching ambition.

3. Solidarity with the Living and the Dead

- The rebel's concerns extend beyond individual existence to encompass all of humanity, including those who have suffered and died. Rebellion is an act of solidarity, a way of honoring the dignity of both the living and the dead.

2. The Camusian Rebel and Death

Death, in Camus's philosophy, is the ultimate absurdity—an inevitable end that renders all human endeavors finite. Yet the rebel does not accept death passively; instead, they confront it as part of their struggle to affirm life.

1. Rebellion Against Death

- For the rebel, the inevitability of death is a challenge to be faced rather than an excuse for resignation. While death cannot be defeated in an absolute sense, the rebel's defiance lies in creating meaning and solidarity in the face of mortality.
- The rebel honors the memory of the dead by striving for justice and by refusing to allow their suffering to be forgotten.

2. The Dignity of the Dead

- Camusian rebellion includes a profound respect for the dead, who are part of the human community. The dead, though silent, bear witness to the injustices and struggles of history, and the rebel seeks to preserve their dignity through remembrance and action.

3. Speculative View of the Rebel on Universal Scientific Resurrection

Would the Camusian rebel embrace the idea of universal scientific resurrection? This speculative exploration suggests that the rebel's response would be complex, reflecting both enthusiasm and caution.

1. The Rebel's Affirmation of Resurrection as a Defiance of Death

- The rebel might view universal scientific resurrection as the ultimate act of defiance against the absurdity of death. By restoring those who have died, humanity would symbolically reject the finality of mortality and affirm the value of every individual life.
- Resurrection could be seen as an extension of human solidarity, a way of honoring the dead by repairing the injustices of the past and offering them a second chance at life.

2. Resurrection as a Collective Act of Solidarity

- The rebel would likely emphasize the communal nature of resurrection, rejecting any vision of it as an individualistic or elitist endeavor. Universal resurrection must serve the collective good, ensuring that all who have died—regardless of status or circumstance—are included.
- In this sense, the rebel might align with the ethical principles underlying universal resurrection, which seek to unite humanity across time and affirm the interconnectedness of all lives.

3. Caution Toward Hubris and Overreach

- The rebel, while supportive of the ideals of resurrection, would also warn against the dangers of hubris. Resurrection, as an act of immense technological and philosophical ambition, risks overstepping the limits of human understanding and power.
- The rebel would urge humanity to proceed with humility, recognizing the ethical complexities and unintended consequences that may arise from attempting such a profound transformation.

4. The Role of Memory and Justice

- The rebel might argue that resurrection should not merely restore life but also seek to address the injustices of the past. For the rebel, resurrection must be an act of justice, ensuring that the resurrected are not merely brought back to life but are given the opportunity to live with dignity and meaning.

- Resurrection should also preserve the memory of the past, honoring the struggles and sacrifices of those who came before. The rebel would caution against any attempt to erase or rewrite history in the process of resurrection.

4. A Fictional Dialogue: The Camusian Rebel Speaks

To further explore this speculative view, imagine the Camusian rebel addressing the concept of universal scientific resurrection in their own words:

"I see in your vision of resurrection a profound act of rebellion—a refusal to accept the finality of death and an affirmation of the value of every human life. To restore the dead is to say that no life is meaningless, that no struggle is forgotten, and that humanity's solidarity extends beyond the boundaries of time.

But I warn you: be wary of hubris. Do not mistake technological power for moral wisdom. Resurrection must not become an act of domination, of imposing your will on the past. It must be an act of humility and justice, a way of honoring the dead and repairing the wrongs of history.

And remember this: resurrection is not enough. To restore life is not to restore meaning. You must ensure that those who are resurrected find a world worth living in—a world of dignity, solidarity, and hope. For without these, resurrection is but an empty gesture, a defiance without purpose.

So rebel wisely, my friends. Let your rebellion against death be a rebellion for life, for justice, for humanity. Only then will your vision of resurrection be worthy of the name."

5. Conclusion: The Rebel's Legacy and Resurrection

The Camusian rebel, as a philosophical archetype, embodies the spirit of defiance, humility, and solidarity that underpins the vision of universal scientific resurrection. While the rebel would embrace the idea of restoring the dead as an affirmation of life and justice, they would also caution against the dangers of hubris and the ethical complexities of such an undertaking.

Through the lens of the rebel, universal scientific resurrection becomes not only a technological challenge but also a profound act of rebellion against the absurd, a way of honoring humanity's shared struggle and building a future that affirms the dignity of all. In this sense, the rebel's voice offers both inspiration and guidance for those who seek to repair the past and transcend the boundaries of mortality.

◊◊◊◊◊

Take Three: A Fictionalized Albert Camus on USR.

Yes, we might reasonably say that the "rebel" man in Albert Camus's The Rebel could also be described as the "meridian" man, as the concept of the "meridian" plays a significant role in Camus's philosophical framework. The meridian represents balance, limits, and the pursuit of justice within the boundaries of human nature. For Camus, the rebel is not merely someone who resists or defies but someone who does so while adhering to the principles of moderation and solidarity—qualities closely tied to the idea of the meridian.

1. The Meridian: A Concept of Balance and Limits

In The Rebel, Camus introduces the idea of the "meridian" as a metaphor for the proper limits of human action and thought. The meridian invokes the image of a line or boundary, symbolizing the point where action and restraint meet—where rebellion is tempered by humility and a recognition of human finitude.

1. Rebellion Without Nihilism

 - The rebel rejects both nihilism, which denies all values, and absolute ideologies, which impose rigid systems of thought. Instead, the rebel seeks a "middle path," affirming the value of life and justice without resorting to tyranny or destruction.

- The meridian represents this balance, ensuring that rebellion does not degenerate into excess or violence but remains rooted in solidarity and respect for human dignity.

2. The Meridian as a Human Scale

- For Camus, the meridian serves as a reminder of humanity's limits: "Every rebellion that transcends its limits prepares a new slavery." The rebel must remain conscious of what is humanly achievable, resisting the temptation to overreach or impose utopian visions that disregard individual freedom.
- The meridian thus reflects the rebel's commitment to a "human scale"—a way of acting that respects the complexities of human existence while striving for justice and meaning.

2. The Rebel as the Meridian Man

The figure of the "rebel" man, as described by Camus, aligns closely with the qualities of the "meridian" man—someone who acts with balance, humility, and a sense of measured defiance.

1. Rebellion and Balance

- The rebel's defiance is not an act of pure destruction but an assertion of values and solidarity. Like the meridian, the rebel seeks to balance resistance with creation, rejecting extremes and embracing moderation.
- For instance, the rebel opposes oppression but does not become an oppressor. They reject injustice while remaining committed to justice that is equitable and humane.

2. Limits and Responsibility

- The meridian man, like the rebel, acknowledges human limits. Camus argues that rebellion must remain within the bounds of what is just and possible, avoiding the pitfalls of absolute ideologies or unrestrained ambition.
- The rebel's actions are guided by a sense of responsibility to others and to the broader human community. This sense of limits

is central to the meridian, which represents a line that should not be crossed.

3. Solidarity and Moderation

- The rebel's solidarity with others mirrors the meridian's emphasis on interconnectedness. The rebel's defiance is not for personal gain but for the collective good, affirming a shared humanity that transcends individual struggles.
- Moderation, a key aspect of the meridian, is also central to the rebel's ethos. The rebel seeks to create meaning and justice without resorting to extremism or absolutism.

3. Camus's Philosophical Vision: The Rebel as a Meridian Archetype

In The Rebel, Camus is deeply concerned with the tension between rebellion and tyranny, creation and destruction. The meridian serves as a philosophical guidepost, ensuring that rebellion remains constructive and life-affirming rather than destructive or nihilistic.

1. Rebellion and Creation

- Camus emphasizes that true rebellion is creative rather than purely destructive. The rebel, like the meridian man, seeks to build a better world without imposing rigid systems or sacrificing individual freedom.
- The rebel's creativity is rooted in balance, reflecting the meridian's principle of moderation and harmony.

2. The Danger of Overstepping

- For Camus, rebellion must remain within the bounds of the meridian to avoid becoming a new form of oppression. When rebellion oversteps these limits—when it seeks to impose absolute ideologies or deny the complexity of human life—it risks descending into tyranny.

- The meridian man embodies the caution and humility necessary to prevent rebellion from becoming destructive or dehumanizing.

3. A Human-Centered Philosophy

- Both the rebel and the meridian man place humanity at the center of their philosophy. They reject both nihilism and transcendental ideologies, focusing instead on the here and now—the tangible realities of human suffering, dignity, and solidarity.
- The meridian represents a commitment to the human scale, ensuring that rebellion serves humanity rather than abstract ideals.

4. Conclusion: The Rebel as the Meridian Man

Based on Camus's The Rebel, it is reasonable to describe the "rebel" man as the "meridian" man, as both figures embody the principles of balance, limits, and solidarity. The rebel's defiance is tempered by humility and a recognition of human finitude, reflecting the meridian's emphasis on moderation and the human scale.

For Camus, the meridian serves as a guide for rebellion, ensuring that it remains constructive and life-affirming rather than destructive or oppressive. In this sense, the rebel and the meridian man are two sides of the same philosophical archetype—symbols of humanity's struggle to find meaning and justice in a world marked by absurdity and suffering.

◊◊◊◊◊

Take Four: A Fictionalized Albert Camus on USR.

The concept of the "Meridian Rebel"—a figure inspired by Albert Camus's The Rebel and centered on balance, limits, and solidarity—offers profound insights that could meaningfully shape and improve the Research and Development (R&D) Program for universal scientific resurrection. By incorporating the spirit of the Meridian Rebel, the program could maintain a sense of ethical responsibility, humility, and justice while pursuing its ambitious vision of resurrecting the dead. This approach would ensure that the program remains grounded in the values of humanity and solidarity, avoiding the pitfalls of hubris, extremism, or dehumanization.

Here's how the spirit of the Meridian Rebel could guide and improve the program:

1. Emphasizing Limits: Proceeding with Humility and Responsibility

The Meridian Rebel recognizes the importance of limits—both in human understanding and in the ethical boundaries of action. Applying this principle to the R&D Program for universal scientific resurrection would encourage researchers to approach the project with a sense of humility and responsibility, acknowledging the complexity and risks involved.

1. Acknowledging the Complexity of Resurrection

   - Resurrection is not merely a technical challenge but a profound intervention in the human condition. The Meridian Rebel would urge researchers to proceed with caution, understanding that the restoration of life involves not just biology but identity, memory, and cultural context.
   - Researchers should remain open to the possibility that some aspects of resurrection may remain unsolvable or require significant philosophical reflection.

2. Avoiding Technological Hubris

   - The spirit of the Meridian Rebel calls for rejecting hubris—the belief that humanity can control or perfect everything through technology. While striving for progress, the program must recognize the potential for unintended consequences, such as ethical dilemmas, misuse of resurrection technologies, or harm to resurrected individuals.

- A commitment to humility would involve ongoing ethical oversight, public accountability, and a willingness to adapt or halt the program if it strays from its ethical foundations.

3. Setting Realistic Goals

- The program should be divided into measured, achievable phases, avoiding overambitious leaps that could compromise safety or ethical integrity. The Meridian Rebel would advocate for steady, incremental progress that allows for reflection and adjustment at each stage.

2. Balancing Ambition with Moderation

The Meridian Rebel seeks a balance between defiance and moderation, resisting both passivity and overreach. For the R&D Program, this balance translates to pursuing the bold vision of universal resurrection while remaining grounded in ethical and practical realities.

1. Avoiding Extremes

- The program should avoid extremes of thought, such as an overemphasis on technological determinism (the belief that technology alone will solve all problems) or an exclusive focus on one method of resurrection at the expense of others.
- Multiple approaches—biological restoration, digital reconstruction, and cosmological methods—should be explored simultaneously, with a willingness to pivot based on evidence and outcomes.

2. Creating a World Worth Living In

- Resurrection is not an end in itself; it must be part of a broader vision of justice, solidarity, and dignity. The Meridian Rebel would remind researchers that resurrected individuals must not only be brought back to life but also integrated into a world where their existence is meaningful and supported.

- This means addressing questions of resource distribution, environmental sustainability, and social equity as part of the program's long-term planning.

3. Ensuring Ethical Oversight

- Moderation also requires robust ethical oversight to ensure that the program remains aligned with its values. The Meridian Rebel would advocate for the establishment of ethical review boards composed of scientists, philosophers, ethicists, and representatives of the global public.
- These boards would be tasked with evaluating the program's progress and ensuring that its methods and goals remain just and humane.

3. Strengthening Solidarity: A Program for All Humanity

The Meridian Rebel's philosophy is rooted in solidarity—the recognition of shared humanity and the commitment to collective justice. Applying this principle to the R&D Program would ensure that universal scientific resurrection serves the common good, rather than narrow interests or elitist ambitions.

1. Global Collaboration

- The program must be a collective effort, involving diverse perspectives from across the globe. This includes not only scientists and researchers but also historians, sociologists, ethicists, and representatives of different cultures and traditions.
- Solidarity also means ensuring that the benefits of resurrection are available to all, regardless of socioeconomic status, geographic location, or historical context. The program must actively work to prevent the exclusion of marginalized groups or the prioritization of certain individuals over others.

2. Honoring the Dignity of the Dead

- The Meridian Rebel would insist that the program treat the dead with dignity and respect, recognizing them as individuals

with histories, identities, and intrinsic value. Resurrection must not reduce individuals to mere data points or biological specimens.
- Efforts should include preserving the cultural, historical, and personal contexts of those being resurrected, ensuring that they are restored as whole persons rather than fragments of their former selves.

3. Engaging the Public

- Solidarity also involves engaging the broader public in the program's development. The Meridian Rebel would call for transparency and dialogue, ensuring that the program reflects the values and concerns of humanity as a whole.
- Public education initiatives, open forums, and participatory decision-making processes could help build trust and foster a sense of shared ownership in the project.

4. Fostering Ethical Creativity: Rebellion as a Force for Justice

For the Meridian Rebel, rebellion is not merely an act of defiance but a creative force that seeks to build a better world. The R&D Program should embody this spirit of creative rebellion, using science and technology to repair the injustices of the past and affirm the value of every human life.

1. A Commitment to Justice

- Resurrection should be framed as an act of justice—an effort to address the suffering and loss experienced by countless individuals throughout history. The Meridian Rebel would remind researchers that the program's ultimate goal is not simply to restore life but to create a world that honors the dignity and worth of every individual.

2. Innovation with Purpose

- The program should encourage innovation that aligns with its ethical foundations. Technologies developed for resurrection—such as advanced medical treatments, artificial intelligence, and space habitats—should also benefit the living, contributing to a more just and sustainable society.

3. Repairing the Past, Not Rewriting It

- The spirit of the Meridian Rebel would caution against attempts to rewrite or erase history through resurrection. Instead, the program should seek to repair the past, acknowledging historical injustices while preserving the integrity of human memory.

5. Conclusion: The Meridian Rebel as a Guiding Spirit

The Meridian Rebel offers a powerful framework for improving the R&D Program for universal scientific resurrection. By emphasizing limits, balance, solidarity, and ethical creativity, the spirit of the Meridian Rebel ensures that the program remains grounded in the values of humanity and justice.

Incorporating these principles would transform the program from a purely technological endeavor into a profound act of rebellion against death and injustice—a project that not only restores life but affirms its meaning and dignity. In this sense, the Meridian Rebel serves as both a guide and an inspiration, reminding us that even the most ambitious visions must remain rooted in the principles of solidarity, responsibility, and hope.

Appendix H.
A Fictionalized Jesus of Nazareth on USR.

The Sermon on the Mount (Matthew 5–7), one of the most profound teachings of Jesus, emphasizes humility, compassion, love, justice, mercy, and spiritual integrity. These principles can serve as a moral and spiritual guide to improve your R&D Program for universal scientific resurrection. By grounding the project in the ethical and spiritual wisdom of the Sermon on the Mount, the program can become not just a technological achievement but a profound reflection of humanity's higher values.

Here's how the teachings of the Sermon on the Mount might inform and inspire meaningful improvements to the resurrection project:

1. The Beatitudes: Guiding the Spirit of the Project

The Beatitudes (Matthew 5:3–12) express blessings for those who embody humility, compassion, and a commitment to justice. These values can shape the ethos of the resurrection project, ensuring that its purpose and methods align with the highest moral principles.

1. Blessed Are the Poor in Spirit (Matthew 5:3):

Humility and Reverence

- The program must approach the act of resurrection with humility, recognizing the enormity of the task and its sacred nature. Bringing people back to life is not just a scientific achievement but a deeply spiritual and ethical act that requires reverence for the mystery of life and death.
- Researchers and leaders involved in the project should cultivate an attitude of service and humility, avoiding pride, arrogance, or the pursuit of fame.

2. Blessed Are They That Mourn (Matthew 5:4):

Healing Grief

- Resurrection must address humanity's shared grief and loss, offering hope and healing to those who mourn for lost loved ones. This means designing the program not only to restore life but also to care for the emotional and spiritual needs of both the living and the resurrected.
- The program could include resources for counseling, spiritual reflection, and support systems to help individuals navigate the profound emotions surrounding resurrection.

3. Blessed Are the Meek (Matthew 5:5):

Gentle Stewardship

- The meek are those who act with gentleness, restraint, and respect. The project should embody this spirit by treating the lives it restores with care and dignity, avoiding any actions that reduce individuals to objects or data points.
- The program must also steward its resources wisely, ensuring that resurrection is pursued in a way that respects the natural world and does not harm future generations.

4. Blessed Are Those Who Hunger and Thirst for Righteousness (Matthew 5:6):

Commitment to Justice

- The project must be a force for justice, ensuring that the benefits of resurrection are distributed equitably and inclusively. Resurrection should not become a privilege for the wealthy or powerful but a universal act of love and fairness.
- Policies and practices must reflect a commitment to social and historical justice, addressing inequities and ensuring that all people—regardless of status, race, or era—are included in the resurrection effort.

5. Blessed Are the Merciful (Matthew 5:7):

Compassion for the Resurrected

- Mercy calls for compassionate care for the resurrected, recognizing that returning to life after death may bring challenges of adjustment, identity, and meaning. The program should provide ongoing emotional, social, and spiritual support for the resurrected, ensuring that they are treated with kindness and understanding.
- For example, the program could include initiatives to help the resurrected reintegrate into society, rediscover purpose, and heal from past traumas.

6. Blessed Are the Peacemakers (Matthew 5:9):

Fostering Harmony

- Resurrection has the potential to unite humanity across time, but it could also create tensions between the living and the resurrected. The program must actively work to foster peace and harmony, building bridges between generations and cultures.
- This could involve creating spaces for dialogue, understanding, and reconciliation, ensuring that the program brings humanity together rather than dividing it.

2. The Call to Be Salt and Light: Inspiring a Higher Purpose

Jesus calls his followers to be the "salt of the earth" and the "light of the world" (Matthew 5:13–16), meaning they should act as agents of preservation, healing, and inspiration. The resurrection project can embody this call by serving as a beacon of hope and a testament to humanity's love and creativity.

1. Preserving the Dignity of Life

- As "salt," the program must work to preserve the dignity and sanctity of life, ensuring that resurrection is pursued with respect for the intrinsic value of every individual.

- This means avoiding any practices that commodify life or treat resurrection as a purely mechanical process detached from its deeper ethical and spiritual implications.

2. Shining a Light of Hope

- As "light," the program can inspire humanity by demonstrating that love and solidarity can transcend even death. Resurrection should not be framed merely as a technological achievement but as a profound act of hope, compassion, and service.
- Publicly communicating the program's ethical and spiritual mission could help foster a sense of shared purpose and global unity.

3. The Golden Rule: Do Unto Others

Jesus teaches, "So in everything, do to others what you would have them do to you" (Matthew 7:12). This ethic of reciprocity is foundational to the Sermon on the Mount and provides a clear guide for the resurrection project.

1. Treating the Resurrected as We Would Want to Be Treated

- The program must prioritize the well-being and dignity of the resurrected, ensuring that they are treated as full human beings rather than as experiments, curiosities, or tools for advancing science.
- This includes respecting their autonomy, providing opportunities for self-expression and purpose, and addressing any needs or concerns they may have.

2. Acting with Empathy and Compassion

- By putting ourselves in the place of the resurrected, we can better understand their potential challenges and needs, such as adjusting to a new world, reconnecting with loved ones, or finding meaning in their return.

- The program's policies and practices should reflect this empathetic perspective, ensuring that every decision is guided by love and care.

4. Loving Your Enemies: Extending Love to All

Jesus's radical teaching to "love your enemies and pray for those who persecute you" (Matthew 5:44) challenges us to act with unconditional love and forgiveness. The resurrection project can embody this teaching by applying universal love and inclusion to its mission.

1. Resurrecting All People, Without Distinction

- The program must ensure that resurrection is offered to all, including those who were historically marginalized, misunderstood, or even considered enemies. Love, as Jesus teaches, is not limited to friends or allies—it extends to all humanity.
- This universal approach would affirm the equal worth of every person, reflecting the radical inclusivity of Jesus's teachings.

2. Healing Historical Divisions

- Resurrection presents an opportunity to reconcile historical conflicts, fostering forgiveness and understanding across time. The program could include initiatives to address the wounds of the past and promote healing and unity among the living and the resurrected.

5. Storing Treasures in Heaven: Focusing on Eternal Values

Jesus warns against storing up treasures on earth, urging his followers to focus on eternal treasures instead (Matthew 6:19–21). This teaching reminds the program to prioritize spiritual and ethical values over material or superficial goals.

1. Pursuing Resurrection as an Act of Love, Not Power

- The program must remain focused on its higher purpose—restoring life as an act of love and service—rather than being driven by the pursuit of fame, wealth, or technological dominance.
- Transparency, humility, and ethical reflection can help ensure that the program stays true to its mission.

2. Valuing People Over Progress

- The program should prioritize the dignity and well-being of individuals over abstract measures of progress or success. This means placing love, compassion, and justice at the center of every decision.

6. Building on the Rock: A Strong Ethical Foundation

Jesus concludes the Sermon on the Mount by urging his followers to build their lives on a solid foundation of his teachings (Matthew 7:24–27). The resurrection project must similarly be built on a strong ethical and spiritual foundation, ensuring that it remains steadfast in the face of challenges.

1. Rooting the Program in Love and Justice

- By grounding every aspect of the project in the principles of the Sermon on the Mount, the program can ensure that its work reflects humanity's highest values.
- Ongoing ethical oversight, spiritual reflection, and dialogue with diverse communities can help maintain this foundation.

Conclusion: Resurrection as a Reflection of the Kingdom of God

The Sermon on the Mount challenges us to embody the values of the Kingdom of God: love, mercy, justice, humility, and hope. By infusing the resurrection project with these principles, it can become a profound act of service that reflects humanity's deepest aspirations and highest callings. Resurrection, in this light, is not just the restoration of life—it is the manifestation of love that transcends death, a testament to the power of compassion, solidarity, and faith.

Appendix I.
A Fictionalized James Joyce on USR.

The artistic spirit of James Joyce's A Portrait of the Artist as a Young Man—centered on individuality, creativity, and the quest for personal and universal meaning—offers a compelling lens through which to rethink and improve the Research and Development (R&D) Program for universal scientific resurrection. Joyce's novel explores themes of artistic rebellion, self-discovery, and the role of imagination in transcending conventional boundaries. These themes could inform the program by infusing it with a sense of creativity, individuality, and profound attention to the human experience, ensuring that the project remains not just a technological endeavor but also an artistic and meaningful act of human expression.

Here's how the artistic spirit of A Portrait of the Artist as a Young Man could inspire and improve the program:

1. Embracing the Individuality of the Resurrected

One of the core themes of Joyce's novel is Stephen Dedalus's journey toward self-realization, where he asserts his individuality against societal conventions. In the same way, the R&D Program must ensure that universal scientific resurrection respects and celebrates the individuality of each resurrected person, rather than reducing them to standardized or generic versions of themselves.

1. Restoring the Unique Essence of Individuals

   - The program must strive to restore not just biological bodies but also the personal identities, memories, and inner lives of resurrected individuals. Just as Stephen seeks to affirm his unique artistic voice, the program must ensure that every resurrected individual reflects the singularity of their personality, experiences, and cultural context.
   - This could involve developing advanced AI and data reconstruction technologies that can integrate genetic information

with historical, cultural, and personal data to recreate not only the physical body but also the subjective essence of the person.

2. Preserving Cultural and Historical Diversity

- Just as Joyce's novel reflects the rich cultural and linguistic tapestry of Ireland, the program should embrace and preserve the diversity of human cultures, languages, and experiences. Resurrection must not erase the cultural specificity of individuals but instead honor the historical and social contexts that shaped their identities.
- This could involve collaborating with historians, anthropologists, and artists to recreate the environments and traditions that were integral to the lives of the resurrected.

2. Infusing the Program with Creative Vision

Stephen Dedalus's declaration—"I will try to express myself in some mode of life or art as freely as I can"—highlights the central role of creativity and freedom of expression in human existence. The R&D Program would benefit from embracing an equally artistic and imaginative approach, viewing resurrection as not just a scientific process but a creative act that seeks to reimagine and repair the human story.

1. Resurrection as an Artistic Act

- Resurrection could be framed as a form of art—an act of creation and storytelling that seeks to reconstruct the intricate tapestry of human lives. By treating resurrection as an artistic endeavor, the program could aim not only for scientific precision but also for emotional depth and meaning.
- This perspective would encourage researchers to think beyond technical challenges and consider the emotional and existential impact of resurrection on both the resurrected individuals and the living.

2. Incorporating Artistic Disciplines into the Program

- Just as Joyce's novel integrates elements of language, mythology, and aesthetics, the program could benefit from collaboration with artists, writers, and philosophers who can contribute imaginative and humanistic perspectives.
- For example, artists could help design simulations or environments that recreate the cultural and emotional worlds of the resurrected, while writers could contribute to the narrative and ethical dimensions of the project.

3. Encouraging Creative Thinking in Science

- The artistic spirit of A Portrait of the Artist as a Young Man could inspire researchers to think creatively and push the boundaries of conventional science. By embracing a mindset of exploration and experimentation, the program could foster innovative solutions to the challenges of resurrection, such as identity reconstruction, memory integration, and ethical dilemmas.

3. Centering the Human Experience

Joyce's novel is deeply concerned with the subjective experience of being human, capturing the thoughts, emotions, and inner conflicts of Stephen Dedalus in rich detail. The R&D Program could adopt a similarly human-centered approach, ensuring that resurrection prioritizes the lived experience of individuals rather than reducing them to abstract data or biological machines.

1. Recreating the Inner Life of the Resurrected

- Resurrection must go beyond physical restoration to address the emotional, psychological, and spiritual dimensions of human existence. Technologies for memory and identity reconstruction should aim to restore not only factual memories but also the emotional textures and subjective experiences that define a person's inner life.

- The program could draw on insights from neuroscience, psychology, and art to better understand and recreate the complexities of human consciousness.

2. Designing Meaningful Lives for the Resurrected

- Stephen Dedalus's struggle to find meaning in his life reflects a universal human need for purpose and self-expression. The program must consider how resurrected individuals will find meaning and fulfillment in their new lives.
- This could involve designing social, cultural, and creative opportunities for the resurrected, ensuring that they have the freedom to explore their identities and contribute to the world in meaningful ways.

4. Balancing Rebellion and Responsibility

Stephen's journey in A Portrait of the Artist as a Young Man is one of rebellion against authority and tradition, as he seeks to forge his own path. Similarly, the R&D Program represents a rebellion against the finality of death and the constraints of the natural world. However, the artistic spirit of the novel also reminds us that rebellion must be tempered by responsibility and ethical reflection.

1. Challenging the Status Quo

- The program's ambition to transcend death aligns with Stephen's rebellious spirit, which challenges conventional notions of religion, morality, and identity. Like Stephen, the program should dare to question existing paradigms and explore new possibilities for human existence.

2. Balancing Individual Freedom with Collective Responsibility

- While Stephen's journey is deeply personal, it is also connected to broader questions of culture and society. Similarly, the program must balance its focus on individual resurrection with a commitment to collective well-being and justice.

- This means ensuring that resurrection serves the common good and does not perpetuate inequality, exploitation, or harm.

5. Conclusion: Resurrection as a Creative and Humanistic Endeavor

The artistic spirit of <u>A Portrait of the Artist as a Young Man</u> offers a powerful framework for improving the R&D Program for universal scientific resurrection. By embracing individuality, creativity, and the richness of the human experience, the program can transcend its technical focus and become a profound act of artistic and humanistic creation.

Resurrection, in this sense, is not merely a scientific process but a <u>celebration of life's complexity and beauty</u>—a way of affirming the unique stories, emotions, and identities that define us as human beings. Like Stephen Dedalus, the program can declare its intention to "forge in the smithy of [its] soul the uncreated conscience of [its] race," using science and art to repair the past and build a future that honors the fullness of human existence.

Appendix J.
A Fictionalized Soren Kierkegaard on USR.

The spirit expressed in Søren Kierkegaard's Works of Love—a deeply Christian philosophical work that emphasizes selfless love, duty, and the eternal value of every human being—provides a profound ethical and spiritual framework to improve the R&D Program for universal scientific resurrection. Kierkegaard's focus on love as the foundation of human existence and his insistence on acting with humility, responsibility, and reverence for the "other" offer essential principles for ensuring that the program remains rooted in compassion, justice, and a sense of the divine.

Here's how the spirit of Works of Love could guide and enhance the R&D Program:

1. Grounding the Program in Selfless Love (Agape)

In Works of Love, Kierkegaard describes love as a commandment and emphasizes its selfless, unconditional, and universal nature. The R&D Program could be improved by aligning its goals and methods with this spirit of agape, prioritizing love for humanity as the guiding principle of resurrection.

1. Resurrection as an Act of Love

   - The program should view resurrection not as an act of scientific triumph or technological progress but as an expression of unconditional love for all who have lived. It is an act that affirms the infinite worth of every person, regardless of their achievements, status, or circumstances.
   - The program must ensure that this act of love extends equally to all, avoiding favoritism or discrimination in deciding who is resurrected.

2. Sacrificial Dedication to the Common Good

   - Kierkegaard emphasizes that true love involves sacrifice and devotion to others. The program can embody this spirit by committing to the <u>well-being of the resurrected</u>, even at great effort or cost. This means prioritizing the needs of individuals over efficiency or convenience, ensuring that resurrection is conducted with care, respect, and reverence for the dignity of each person.

3. <u>Universal Inclusion</u>

- Love, as Kierkegaard describes it, is universal and indiscriminate. In this spirit, the program must aim to resurrect all who have ever lived, transcending distinctions of race, gender, wealth, or social status. Every individual must be treated as equally valuable, reflecting the divine love that Kierkegaard insists is at the heart of human relationships.

2. Recognizing the Eternal Value of Each Person

Kierkegaard writes that every human being has an <u>eternal significance</u> because they are loved by God. This perspective calls the R&D Program to honor the <u>infinite worth and uniqueness of every individual</u> being resurrected.

1. <u>Restoring Not Only Life but Identity</u>

- Resurrection must not reduce people to biological or digital entities but must seek to restore the <u>fullness of their identity</u>, including their memories, relationships, and individuality.
- The program should develop methods that honor the uniqueness of each person, ensuring that resurrection is not a mechanical process but one that respects the <u>sacred mystery of personhood</u>.

2. <u>Reverence for the Divine Image in Humanity</u>

- Kierkegaard's theology emphasizes that humans are created in the image of God and are therefore sacred. The program must treat the process of resurrection with <u>reverence and humility</u>, recognizing that it is engaging with something far greater than mere science—it is participating in the restoration of beings with eternal significance.

- 322 -

- This could involve incorporating spiritual and philosophical reflection into the program, ensuring that the ethical and metaphysical dimensions of resurrection are not neglected.

3. Infusing Ethical Responsibility into Every Stage of the Program

Kierkegaard emphasizes that love is not merely an emotion but a duty—a commandment that calls us to act ethically and responsibly toward others. The R&D Program could benefit from this principle by ensuring that every stage of its development is guided by ethical responsibility and a commitment to the well-being of both the resurrected and the living.

1. Responsibility Toward the Resurrected

- The program must ensure that resurrected individuals are treated with care, dignity, and respect. This includes addressing their emotional, psychological, and social needs, ensuring that they are not merely revived but are given the opportunity to live meaningful and fulfilling lives.
- Kierkegaard's emphasis on love as an act of service calls the program to serve the resurrected, placing their needs and well-being at the center of its mission.

2. Responsibility Toward the Living

- While the program focuses on resurrecting the dead, it must also consider its impact on the living. Kierkegaard's ethic of love requires that the program act in ways that promote solidarity and harmony between the living and the resurrected, avoiding conflicts or divisions.
- This could involve fostering dialogue and understanding between different groups, ensuring that the program unites humanity across time rather than creating new barriers or inequalities.

3. Transparency and Humility

- Kierkegaard's philosophy warns against pride and self-serving motives. The program must operate with humility, avoiding any sense of arrogance or triumphalism about its achievements.
- Transparency in its goals, methods, and outcomes would reflect this humility, ensuring that the program remains accountable to the global community and aligned with its core mission of love and service.

4. Building a Community of Love and Solidarity

Kierkegaard's Works of Love emphasizes the importance of building relationships rooted in love and mutual support. The R&D Program could improve by fostering a culture of love and solidarity among the living, the resurrected, and the global community.

1. Creating a World of Love for the Resurrected

- Resurrection must not bring people back into a world of suffering or alienation. The program should work to create a society founded on love, justice, and mutual care, ensuring that the resurrected are welcomed into a community that affirms their dignity and supports their flourishing.
- This involves addressing systemic issues like inequality, environmental sustainability, and resource distribution, ensuring that the world can sustain and nurture an expanded human population.

2. Fostering Intergenerational Solidarity

- Kierkegaard's vision of love extends across time, emphasizing that love binds together all humanity. The program must foster solidarity between the living and the resurrected, creating opportunities for mutual understanding, learning, and collaboration.
- This could involve designing systems and structures that encourage the resurrected to contribute meaningfully to society,

sharing their unique perspectives and experiences with future generations.

3. Uniting Humanity Through Love

- The program should view resurrection as a way of healing historical divisions and uniting humanity across eras. By resurrecting individuals from all periods and cultures, the program can affirm the interconnectedness of all human lives and create a shared sense of purpose and belonging.

5. Treating Resurrection as a Divine Responsibility

For Kierkegaard, love is ultimately rooted in God, and all human actions must reflect a sense of responsibility to the divine. The R&D Program could draw on this principle by recognizing resurrection as a sacred responsibility, one that requires reverence, humility, and alignment with God's will.

1. Viewing Resurrection as Participation in Divine Love

- The program can frame its work as an act of participating in God's love, seeking to restore life not for human glory but as a reflection of the divine commandment to love others as ourselves.
- This perspective would encourage researchers to approach their work with a sense of awe and reverence, recognizing the sacred nature of the lives they are restoring.

2. Incorporating Spiritual Reflection

- The program could benefit from ongoing spiritual and philosophical reflection, ensuring that its actions remain aligned with the higher calling of love and service.
- This could involve consulting with theologians, ethicists, and spiritual leaders, fostering dialogue about the ethical and metaphysical dimensions of resurrection.

6. Conclusion: Resurrection as an Act of Eternal Love

The spirit of Kierkegaard's <u>Works of Love</u> provides a powerful framework for improving the R&D Program for universal scientific resurrection. By grounding the program in <u>selfless love, reverence for the eternal value of each person, and ethical responsibility</u>, it can transcend its technical goals and become a profound act of service and solidarity.

Resurrection, in this light, is not merely a scientific achievement but an expression of <u>unconditional love for all humanity</u>, a way of affirming the infinite worth of every individual and healing the divisions of history. Inspired by Kierkegaard's vision, the program can become a testament to the power of love to overcome death and create a future rooted in justice, compassion, and eternal hope.

Appendix K.
A Fictionalized Gerard O'Neill on USR.

The vision of <u>universal scientific resurrection</u>—the restoration of all who have died through advanced technology—raises profound questions about the future of humanity and our capacity to sustain a vastly expanded population. If successful, universal scientific resurrection would require not only the ability to restore individuals but also sufficient resources, space, and infrastructure to accommodate billions (or even trillions) of resurrected individuals. In this context, the concept of <u>O'Neill extraterrestrial habitats</u>—self-sustaining space colonies designed to support large human populations—offers a compelling solution to the logistical challenges posed by universal resurrection.

This appendix explores the relationship between O'Neill habitats and universal scientific resurrection, examining how these visionary space-based structures could provide the environmental, economic, and ethical foundation for achieving a future in which humanity transcends death and expands its reach beyond Earth.

1. What Are O'Neill Extraterrestrial Habitats?

O'Neill habitats are large, self-sustaining structures in space designed to support human life indefinitely. They were first proposed by physicist <u>Gerard K. O'Neill</u> in his seminal work <u>The High Frontier: Human Colonies in Space</u> (1976), which outlined a vision for humanity's expansion into space as a means of addressing Earth's limitations.

1. <u>Design and Structure</u>

   - O'Neill habitats are typically envisioned as massive cylindrical or toroidal (donut-shaped) structures located in <u>free space</u>, such as at <u>Lagrange points</u>—stable regions in the Earth-Moon system where gravitational forces balance.
   - These habitats would rotate to produce artificial gravity through centrifugal force, while their interiors would feature

Earth-like environments, including atmospheric regulation, agricultural zones, and residential areas.

2. Self-Sustaining Ecosystems

- O'Neill habitats are designed to be entirely self-sufficient, relying on renewable energy (such as solar power) and closed-loop ecosystems to support human life. Advanced recycling systems would ensure that water, air, and nutrients are continuously replenished.
- Resources for constructing and sustaining these habitats would be sourced from space itself, such as mining asteroids or utilizing materials from the Moon.

3. A Solution to Earth's Limitations

- O'Neill argued that space habitats offered a way to address the challenges of overpopulation, environmental degradation, and resource scarcity on Earth. By expanding outward, humanity could create new frontiers for growth and innovation, ensuring a sustainable future for billions—or even trillions—of people.

2. The Challenges of Universal Scientific Resurrection

Universal scientific resurrection envisions the restoration of all who have died, potentially increasing the human population by orders of magnitude. This ambitious goal presents significant logistical challenges, particularly in terms of space, resources, and sustainability.

1. Population Expansion

- If universal resurrection is achieved, the human population would no longer be limited to the living but would include all who have ever lived. Estimates suggest that over 100 billion people have lived and died throughout history. Accommodating such a vast number of individuals would require a dramatic expansion of humanity's living space and resource base.

2. Earth's Finite Resources

- Earth's ecosystems and resources are already under strain from the current population of 8 billion people. Adding billions more resurrected individuals would place unsustainable pressure on the planet's capacity to provide food, water, energy, and shelter.

3. Ethical Considerations

- The ethical imperative of universal resurrection assumes that all lives are valuable and deserve to be restored. However, this commitment also requires ensuring that resurrected individuals are provided with a dignified existence—one that includes adequate space, resources, and opportunities for meaningful lives.

3. O'Neill Habitats as a Solution

O'Neill habitats offer a practical and visionary framework for addressing the challenges of universal scientific resurrection. By expanding humanity's presence into space, these habitats could provide the infrastructure necessary to support a vastly expanded population while ensuring sustainability and quality of life.

1. Infinite Space and Resources

- Space is effectively infinite, offering boundless opportunities for the construction of new habitats. By utilizing resources from asteroids and other celestial bodies, humanity could create enough living space to accommodate even the largest possible population.
- O'Neill habitats could be constructed in large numbers, each capable of supporting tens of thousands to millions of individuals. Collectively, these habitats could form a vast network of human settlements distributed throughout the solar system.

2. Sustainability and Self-Sufficiency

- The closed-loop ecosystems of O'Neill habitats are designed to operate independently of Earth, relying on renewable energy and space-based resources. This self-sufficiency ensures that humanity's expansion into space would not exacerbate the environmental challenges facing Earth.

- These habitats could also serve as laboratories for developing advanced technologies in recycling, agriculture, and energy production—technologies that could further enhance the feasibility of universal resurrection.

3. A New Frontier for Resurrected Lives

- O'Neill habitats offer more than just physical space—they provide <u>new opportunities for exploration, creativity, and community</u>. Resurrected individuals could contribute to the development and governance of these habitats, participating in a collective effort to shape humanity's future.
- The creation of space-based societies would allow humanity to transcend its terrestrial limitations, offering a chance to build a civilization that reflects the values of justice, solidarity, and hope that underpin the vision of universal resurrection.

4. Practical Steps Toward O'Neill Habitats

While the construction of O'Neill habitats remains a long-term goal, significant progress is being made in the technologies and infrastructure necessary to realize this vision.

1. Space Resource Utilization

- Advances in asteroid mining and lunar resource extraction are laying the groundwork for space-based construction. These resources could provide the raw materials needed to build large-scale habitats, reducing dependence on Earth.

2. Renewable Energy in Space

- Solar power satellites and other space-based energy systems are being developed to provide abundant and renewable energy for space habitats. These technologies are essential for powering the ecosystems and industries within O'Neill habitats.

3. International Collaboration

- The development of O'Neill habitats will require unprecedented levels of international collaboration, bringing together governments, private companies, and scientific institutions to pool resources and expertise.

4. A Philosophy of Expansion

- Beyond technological advancements, the realization of O'Neill habitats requires a cultural shift toward embracing humanity's potential for growth and exploration. This philosophy aligns closely with the values of universal scientific resurrection, emphasizing the interconnectedness of all human lives and the boundless possibilities of the future.

5. Ethical Implications of Space Habitats for Resurrection

The use of O'Neill habitats as a foundation for universal resurrection raises profound ethical questions about humanity's responsibilities to the resurrected and the broader cosmos.

1. Respect for the Resurrected

- O'Neill habitats must ensure that resurrected individuals are treated with dignity and provided with opportunities for meaningful lives. This includes access to education, healthcare, and community—essentials for rebuilding lives in a new context.

2. A New Relationship with the Cosmos

- Expanding into space requires humanity to adopt a new ethical framework that respects both the resources of the cosmos and the needs of future generations. O'Neill habitats should be designed to balance growth with sustainability, ensuring that humanity's expansion does not come at the cost of planetary ecosystems or intergenerational justice.

3. Solidarity Across Time and Space

- The creation of space habitats reflects a commitment to solidarity—not only among the living but also with the resurrected. These habitats serve as a testament to humanity's belief in the value of every life and its determination to create a future in which no one is left behind.

Conclusion

O'Neill extraterrestrial habitats offer a visionary solution to the challenges posed by universal scientific resurrection. By providing sustainable, self-sufficient living spaces in the vast expanse of space, these habitats enable humanity to transcend the limitations of Earth and create a future in which resurrection becomes not only possible but sustainable.

As humanity looks to the cosmos for new frontiers, O'Neill habitats represent a profound opportunity to unite the living and the resurrected in a shared effort to repair the past and build a better future. They are a testament to the boundless potential of human creativity, courage, and hope—a foundation for a world where death is no longer the end, but the beginning of a new chapter in the human story.

Appendix L.
Additional Suggestions And Comments Related to USR.

To improve your resurrection project, I [You.com Smart Assistant] would suggest drawing inspiration from a variety of philosophical, literary, and theological works that explore themes of life, death, love, justice, and the human condition. These works can provide ethical guidance, creative inspiration, and a deeper understanding of the profound implications of resurrection. Here are some suggestions:

1. Mary Shelley's **Frankenstein**

Shelley's Frankenstein is a cautionary tale about the dangers of scientific hubris and the ethical responsibilities of creators toward their creations. While your resurrection project is likely motivated by love and justice, this novel serves as a reminder to approach the endeavor with humility, foresight, and care. It emphasizes the importance of considering the emotional and social needs of those brought back to life, as well as the potential unintended consequences of such a profound act.

Key Takeaways:

- Ensure that the resurrected are treated with dignity and compassion, not as experiments or tools.
- Anticipate and address the psychological and societal challenges that resurrection might create.

2. Viktor Frankl's **Man's Search for Meaning**

Frankl's work, rooted in his experiences as a Holocaust survivor, explores the human need for meaning, even in the face of suffering and death. Resurrection is not just about restoring life but also about helping individuals find purpose and meaning in their new existence. Frankl's insights can guide the program in creating opportunities for the resurrected to live fulfilling and meaningful lives.

Key Takeaways:

- Design systems that help the resurrected rediscover purpose and integrate into society.
- Recognize that meaning is central to human flourishing, and resurrection must address more than physical needs.

3. Dante Alighieri's **Divine Comedy**

Dante's Divine Comedy offers a rich exploration of the afterlife, human morality, and the journey toward redemption and divine love. While your project focuses on physical resurrection, Dante's work can inspire a deeper reflection on the spiritual and ethical dimensions of life and death. It also emphasizes the interconnectedness of all humanity and the importance of love and justice in shaping our actions.

Key Takeaways:

- Consider the spiritual implications of resurrection and how it aligns with humanity's broader quest for redemption and meaning.
- Use the project as an opportunity to promote love, justice, and unity across generations and cultures.

4. Martin Buber's **I and Thou**

Buber's philosophy emphasizes the importance of genuine, reciprocal relationships between individuals. Resurrection must not reduce people to objects or data points but must treat them as unique, sacred beings capable of meaningful relationships. Buber's work can guide the program in fostering authentic connections between the resurrected and the living.

Key Takeaways:

- Treat the resurrected as full persons, not as objects of scientific achievement.
- Foster relationships of mutual respect and understanding between the living and the resurrected.

5. Hannah Arendt's **The Human Condition**

Arendt's exploration of human activity—labor, work, and action—offers valuable insights into the ethical and social dimensions of resurrection. Her concept of <u>natality</u> (the capacity for new beginnings) aligns with the idea of resurrection as a profound act of renewal. Arendt's work can help the program consider how resurrection might reshape human society and relationships.

Key Takeaways:

- View resurrection as an opportunity for renewal and transformation, not just restoration.
- Consider the societal and political implications of resurrection and how it might foster new forms of community and solidarity.

6. The Bhagavad Gita

The <u>Bhagavad Gita</u> offers profound reflections on life, death, duty, and the eternal soul. Its teachings on detachment, selfless action, and the interconnectedness of all beings can provide spiritual guidance for the resurrection project. The Gita emphasizes acting with <u>humility and devotion</u>, aligning well with the ethical responsibilities of such an endeavor.

Key Takeaways:

- Approach resurrection with a sense of sacred duty and selflessness.
- Recognize the interconnectedness of all life and ensure that the project serves the common good.

7. Albert Schweitzer's **Reverence for Life**

Schweitzer's philosophy of "reverence for life" emphasizes the intrinsic value of all living beings and the ethical responsibility to preserve and nurture life. This principle aligns closely with the goals of your resurrection project and can guide its ethical framework.

Key Takeaways:

- Treat resurrection as an act of reverence for the sanctity of life.
- Ensure that the project promotes compassion, care, and respect for all individuals.

8. Fyodor Dostoevsky's **The Brothers Karamazov**

Dostoevsky's novel wrestles with profound questions of faith, suffering, and the human condition. The character of Alyosha Karamazov embodies a Christ-like love and compassion that can inspire the program to prioritize love, forgiveness, and solidarity in its mission.

Key Takeaways:

- Use resurrection as an opportunity to heal historical wounds and foster reconciliation.
- Emphasize love and compassion as the guiding principles of the project.

9. Simone Weil's **Gravity and Grace**

Weil's reflections on suffering, grace, and the human longing for transcendence can provide a spiritual dimension to the resurrection project. Her emphasis on humility and attention to the needs of others aligns with the ethical responsibilities of bringing people back to life.

Key Takeaways:

- Approach resurrection with humility and a sense of service to others.
- Recognize the spiritual and existential dimensions of life and death.

10. The Qur'an

The Qur'an contains numerous reflections on resurrection, justice, and the afterlife. Its teachings on the <u>Day of Resurrection</u> and the accountability of human actions can inspire the program to consider the moral and spiritual implications of its work.

Key Takeaways:

- Reflect on the ethical and spiritual responsibilities of resurrection.
- Ensure that the project aligns with principles of justice, mercy, and compassion.

Conclusion: A Holistic Approach to Resurrection

By drawing on these works, your resurrection project can transcend its technical goals and become a profound act of love, justice, and renewal. These texts remind us that resurrection is not just about restoring life—it is about honoring the dignity, meaning, and interconnectedness of all human beings. By integrating these insights, the project can reflect humanity's highest values and aspirations.

Appendix M.
Cryobiology.

Proposal:
Research and Development Program for Perfecting Long-Term Human Organ Cryopreservation for Transplantation

Title

Project VitalFreeze: Advancing Cryopreservation Science for Long-Term Human Organ Storage and Transplantation

Executive Summary

Organ transplantation is a life-saving procedure, but its success is often limited by the availability of viable donor organs and the short window of time in which organs remain functional outside the human body. Current organ preservation techniques, such as cold storage, are inadequate for long-term storage and result in significant organ wastage.

This R&D program, Project VitalFreeze, aims to perfect the science of long-term human organ cryopreservation, enabling the indefinite storage of donor organs while ensuring they remain functional for transplantation. By leveraging breakthroughs in cryobiology, nanotechnology, and tissue engineering, the program seeks to address critical challenges such as ice formation, tissue damage during freezing and thawing, and cryoprotectant toxicity. Success in this endeavor will revolutionize organ transplantation, dramatically reducing the organ shortage crisis and saving countless lives.

Vision

To develop a robust, scalable, and safe cryopreservation technology that enables long-term storage of human organs for transplantation, ensuring universal availability of viable organs for all patients in need.

Mission

To conduct interdisciplinary research on advanced cryopreservation techniques, from cryoprotectant development to precision thawing methods, while addressing biological, technical, and logistical barriers to organ viability and transplantation success.

1. Objectives

The program will focus on achieving the following key objectives:

Scientific Objectives

1. Prevent Ice Formation: Develop methods to prevent intracellular and extracellular ice crystal formation during freezing, which can destroy organ structure and function.
2. Reduce Cryoprotectant Toxicity: Design and test non-toxic cryoprotectants that protect organs during freezing without causing cellular damage.
3. Perfect Thawing Techniques: Develop precise and uniform thawing methods to prevent thermal stress, ice recrystallization, and tissue damage.
4. Validate Organ Viability: Demonstrate that organs preserved using the technology remain functional and suitable for transplantation.

Technical Objectives

1. Develop Scalable Protocols: Create protocols and equipment for preserving and thawing organs at both clinical and industrial scales.
2. Automate Cryopreservation Processes: Design automated systems for consistent, replicable cryopreservation and thawing of human organs.
3. Optimize Storage Conditions: Research and optimize ultra-low temperature storage environments to maintain long-term organ viability.

Ethical and Societal Objectives

1. Ensure Accessibility: Create cost-effective cryopreservation solutions to make the technology widely available across healthcare systems.
2. Minimize Risks: Ensure the safety and bioethics of the technology through rigorous oversight and testing.
3. Engage Stakeholders: Collaborate with medical professionals, regulatory bodies, and donors to build trust and acceptance of cryopreserved organs.

2. Research Scope

The R&D program will be divided into four main phases:

Phase 1: Foundational Research

Duration: 2 years

- Cryoprotectant Development: Research and develop advanced cryoprotectant solutions capable of vitrification (glass-like freezing) without ice formation or toxicity.
- Biological Studies: Investigate the effects of cryopreservation on organ-specific cell types (e.g., liver hepatocytes, kidney nephrons).
- Nanotechnology Applications: Explore nanomaterials and nanoparticles for delivering cryoprotectants evenly and improving thermal conductivity during freezing and thawing.

Deliverables:

- Initial cryoprotectant formulations with reduced toxicity.
- Proof-of-concept studies demonstrating vitrification of small tissue samples.

Phase 2: Scaling Cryopreservation to Whole Organs

Duration: 3 years

- <u>Organ-Specific Protocols</u>: Adapt and optimize cryopreservation protocols for specific organs, such as kidneys, hearts, and livers, which have unique structural and metabolic requirements.
- <u>Freezing and Thawing Systems</u>: Develop precision-controlled freezing and reheating systems to ensure uniform temperature distribution and prevent tissue damage.
- <u>Organ Viability Testing</u>: Test preserved organs in laboratory settings to assess structural integrity, cellular function, and metabolic activity post-thaw.

<u>Deliverables</u>:

- Successful cryopreservation and thawing of whole organs with minimal tissue damage.
- Data on organ viability and functionality post-preservation.

<u>Phase 3: Preclinical and Clinical Validation</u>

<u>Duration</u>: 4 years

- <u>Animal Studies</u>: Conduct preclinical studies to transplant cryopreserved organs into animal models and monitor outcomes such as graft survival, immune response, and functionality.
- <u>Human Trials</u>: Initiate clinical trials to evaluate the safety and efficacy of cryopreserved organs for transplantation in human patients.
- <u>Regulatory Compliance</u>: Work with regulatory agencies (e.g., FDA, EMA) to ensure compliance with guidelines and obtain approvals for clinical use.

<u>Deliverables</u>:

- Preclinical data demonstrating successful transplantation outcomes.
- Initial human trial results showing viability and safety of cryopreserved organs.

Phase 4: Industrial Application and Global Deployment

Duration: 3 years

- Manufacturing Systems: Develop scalable manufacturing processes for cryopreservation tools and consumables (e.g., cryoprotectants, freezing systems).
- Distribution Networks: Establish global networks for organ storage, distribution, and transportation using cryopreservation technology.
- Training and Guidelines: Train healthcare providers in cryopreservation techniques and develop standardized guidelines for clinical use.

Deliverables:

- Commercially viable cryopreservation systems.
- Deployment of cryopreservation technology in leading transplantation centers.

3. Resources and Budget

Personnel

- Cryobiologists specializing in organ freezing and vitrification.
- Biomedical engineers for developing freezing and thawing systems.
- Physicians and transplant surgeons to provide clinical insights.
- Ethicists and regulatory experts to address societal and legal considerations.

Facilities

- Cryobiology laboratories equipped with ultra-low temperature freezers, vitrification tools, and advanced imaging systems.
- Preclinical research facilities for testing cryopreserved organs in animal models.
- Clinical trial sites for testing in human transplantation settings.

Budget Estimate

Total Estimated Budget: $100 million over 12 years
- Phase 1: $20 million (cryoprotectant development, foundational research)
- Phase 2: $30 million (organ-specific protocols, freezing systems)
- Phase 3: $35 million (preclinical and clinical trials)
- Phase 4: $15 million (industrial scaling, global deployment)

4. Key Challenges and Mitigation Strategies

Scientific Challenges

- Ice Formation: Ice crystals can cause irreversible damage to organs during freezing.
  Mitigation: Focus on vitrification techniques and rapid cooling methods.
- Cryoprotectant Toxicity: High concentrations of cryoprotectants can harm cells.
  Mitigation: Develop non-toxic or biodegradable cryoprotectants.

Technical Challenges

- Uniform Freezing and Thawing: Uneven temperature distribution can lead to localized tissue damage.
  Mitigation: Use nanotechnology and advanced thermal control systems.

Ethical Challenges

- Donor Consent: Ensure clear and informed consent for the use of organs in cryopreservation research.
  Mitigation: Work with ethical review boards to establish transparent policies.

5. Expected Outcomes

- Scientific Advancements: Breakthroughs in cryobiology, organ preservation, and transplantation medicine.

- <u>Organ Banks</u>: Creation of "organ banks" where donor organs can be stored indefinitely, eliminating the need for immediate matching and transportation.
- <u>Reduced Organ Shortages</u>: Increased availability of viable organs for transplantation, potentially saving millions of lives.
- <u>Global Impact</u>: Improved access to organ transplants in underserved regions, reducing health disparities worldwide.

6. Societal and Medical Impacts

- <u>Transforming Transplantation Medicine</u>: Perfect cryopreservation will redefine how organ transplantation is performed, enabling better patient outcomes and reducing organ rejection.
- <u>Emergency Preparedness</u>: Cryopreservation could provide a stockpile of organs for emergencies, such as natural disasters or pandemics.
- <u>Economic Benefits</u>: Long-term storage and improved organ utilization could reduce healthcare costs associated with organ failure and transplantation logistics.

7. Call to Action

We invite collaboration with government agencies, academic institutions, biotech companies, and philanthropic organizations to fund and support this transformative project. Together, we can overcome the challenges of organ cryopreservation and create a future where no patient dies waiting for a transplant.

<u>Contact Information</u>:

[Insert Name/Organization of Proposal Lead]
[Email Address]
[Phone Number]

This proposal outlines a clear, actionable plan to perfect long-term human organ cryopreservation, addressing one of the most pressing challenges in modern medicine. By investing in this initiative, we can save countless lives and revolutionize transplantation science.

Proposed Legislation:
National Institute on Low Temperature Biology Act

Title

National Institute on Low Temperature Biology Act

Preamble

To establish the National Institute on Low Temperature Biology (NILTB) as a federal research institute dedicated to advancing the science of low-temperature biology, cryobiology, and cryopreservation, with the goal of improving human health, agriculture, and environmental sustainability.

Section 1: Short Title

This Act may be cited as the "National Institute on Low Temperature Biology Act."

Section 2: Findings and Purpose

Findings

The Congress finds that:
1. Organ Shortages and Cryopreservation: Over 100,000 individuals in the United States are currently on organ transplant waiting lists, and many die each year due to the lack of viable donor organs. Advances in cryopreservation could enable long-term organ storage, reducing organ shortages and saving lives.
2. Agricultural and Environmental Applications: Low-temperature biology has applications in preserving genetic material for endangered species, improving crop resilience, and mitigating the effects of climate change on ecosystems.
3. Scientific and Economic Potential: The United States has the opportunity to lead the world in cryobiology and low-temperature

research, fostering innovation, creating jobs, and advancing biotechnology.
4. Existing Gaps in Research: Despite the importance of low-temperature biology, there is no centralized federal institute dedicated to advancing this field, leaving critical gaps in research coordination and funding.

Purpose

The purpose of this Act is to:
1. Establish the National Institute on Low Temperature Biology (NILTB) to coordinate and fund research in cryobiology, cryopreservation, and related fields.
2. Promote the development of technologies for long-term organ preservation, tissue engineering, and genetic material storage.
3. Support interdisciplinary research to address challenges in agriculture, medicine, and environmental conservation.
4. Ensure the ethical and equitable application of low-temperature biology technologies.

Section 3: Establishment of the National Institute on Low Temperature Biology

(a) Establishment

There is established within the Department of Health and Human Services (HHS) a new federal research institute to be known as the National Institute on Low Temperature Biology (NILTB).

(b) Mission

The mission of the NILTB shall be to:
1. Conduct and support research on the biological effects of low temperatures on cells, tissues, and organisms.
2. Develop and refine cryopreservation techniques for human organs, tissues, and genetic material.
3. Advance the understanding of low-temperature biology in agriculture, including seed preservation and crop resilience.

4. Promote the preservation of biodiversity through cryogenic storage of genetic material from endangered species.
5. Facilitate the translation of research findings into practical applications in medicine, agriculture, and environmental science.

(c) Headquarters

The NILTB shall be headquartered in a location determined by the Secretary of Health and Human Services, with consideration given to proximity to leading research institutions and biotechnology hubs.

Section 4: Powers and Duties

The NILTB shall have the authority to:
1. Award Grants and Contracts: Provide funding to universities, research institutions, and private entities for projects related to low-temperature biology.
2. Coordinate Research: Collaborate with other federal agencies, including the National Institutes of Health (NIH), the National Science Foundation (NSF), and the Department of Agriculture (USDA), to align research priorities.
3. Establish Research Centers: Create regional centers of excellence in low-temperature biology to foster innovation and collaboration.
4. Develop Standards: Work with regulatory agencies to establish standards for the safe and ethical application of cryopreservation technologies.
5. Promote Public Awareness: Educate the public and stakeholders about the benefits and ethical considerations of low-temperature biology.

Section 5: Governance

(a) Director

The NILTB shall be headed by a Director, who shall be appointed by the President with the advice and consent of the Senate. The Director shall:

1. Serve a term of 5 years, with the possibility of reappointment.
2. Oversee the operations, budget, and strategic direction of the NILTB.
3. Report annually to Congress on the Institute's activities, achievements, and challenges.

(b) Advisory Board

The NILTB shall establish an Advisory Board composed of:
1. Experts in cryobiology, biotechnology, medicine, agriculture, and environmental science.
2. Representatives from federal agencies, academia, and industry.
3. Ethicists and public representatives to ensure diverse perspectives.

The Advisory Board shall provide guidance on research priorities, ethical considerations, and public engagement.

Section 6: Funding

(a) Authorization of Appropriations

There is authorized to be appropriated to the NILTB:
1. $500 million for the first fiscal year following the enactment of this Act.
2. $2 billion over the subsequent 5 fiscal years, with annual adjustments for inflation.

(b) Allocation of Funds

Funds appropriated under this Act shall be allocated as follows:
1. 50% for competitive research grants and contracts.
2. 25% for the establishment and operation of regional research centers.
3. 15% for public education and outreach programs.
4. 10% for administrative and operational expenses.

Section 7: Ethical Considerations

The NILTB shall:
1. Develop and enforce ethical guidelines for research involving human and animal tissues.
2. Ensure that cryopreservation technologies are accessible and equitable, particularly for underserved communities.
3. Collaborate with international organizations to promote global standards for low-temperature biology research.

Section 8: Reporting and Accountability

(a) Annual Report

The NILTB shall submit an annual report to Congress detailing:
1. Research progress and breakthroughs.
2. Allocation and use of funds.
3. Challenges and recommendations for future action.

(b) Independent Review

Every 5 years, the NILTB shall undergo an independent review by a panel of experts to evaluate its effectiveness and recommend improvements.

Section 9: Effective Date

This Act shall take effect 90 days after the date of its enactment.

Section 10: Severability

If any provision of this Act, or the application thereof, is held invalid, the remainder of the Act and the application of such provisions to other persons or circumstances shall not be affected.

Conclusion

The establishment of the National Institute on Low Temperature Biology will position the United States as a global leader in cryobiology and low-temperature research. By addressing critical challenges in organ transplantation, agriculture, and environmental conservation, this legislation will save lives, protect biodiversity, and drive scientific innovation. Congress is urged to act swiftly to pass this transformative legislation.

Appendix N.
Extraterrestrial Governance.

Governance Models for O'Neill Extraterrestrial Habitats

The development of O'Neill habitats—massive space-based settlements designed for living and working in environments superior to Earth—represents a transformative shift in humanity's relationship with space and governance. These habitats will face unique challenges, ranging from resource management in closed systems to the regulation of inter-habitat trade and ensuring human rights in an extraterrestrial context. Below is an outline of several alternative governance models that could be applied to this new world, each with strengths, weaknesses, and suitability depending on the scale and diversity of the habitats.

1. Federalist Model: The "Space Federation"

Overview

This model envisions a federation of O'Neill habitats, each with self-governing autonomy but united under a central governing body for overarching policies, security, and coordination. Habitats would act as "states" or "provinces" within a broader inter-habitat federation.

Key Features

- Decentralized Autonomy: Individual habitats govern their local affairs, including resource distribution, local laws, and cultural practices.
- Central Coordination: A central governing body handles defense, trade regulations, space traffic control, and inter-habitat dispute resolution.
- Constitutional Framework: A founding document outlines rights, responsibilities, and the division of powers between local and central entities.
- Representative Legislature: Each habitat elects representatives to participate in a central legislative assembly.

Strengths

- Balances local autonomy with centralized coordination.
- Encourages diversity and experimentation in governance across habitats.
- Provides a framework for equitable trade and defense coordination.

Weaknesses

- Risk of bureaucratic inefficiency in the central authority.
- Potential for conflicts between local and central governments.
- Requires strong mechanisms for dispute resolution.

Example

An interplanetary "United Federation of Space Habitats" with a rotating capital located in one of the larger, more established habitats.

2. Corporate Governance Model: The "Space Megacorp"

Overview

O'Neill habitats could be established and governed by corporations under a charter system, where the corporation owns and manages the habitat while residents are shareholders, employees, or contractual citizens.

Key Features

- Corporate Ownership: The corporation is responsible for infrastructure, resource management, and governance.
- Contractual Citizenship: Residents agree to abide by the company's regulations and policies in exchange for housing, work, and resources.
- Profit-Driven Governance: Decisions prioritize economic sustainability and growth.

- <u>Meritocratic Leadership</u>: Leaders are appointed based on corporate performance metrics rather than public elections.

<u>Strengths</u>

- Efficient decision-making driven by profit and sustainability incentives.
- Strong focus on economic growth and technological development.
- Simplifies governance in the early stages of habitat establishment.

<u>Weaknesses</u>

- Risk of prioritizing profit over resident well-being or rights.
- Limited democratic participation; residents have little say in governance.
- Potential for exploitation of labor and resources.

<u>Example</u>

A habitat owned by a space manufacturing conglomerate where residents work in high-tech industries (e.g., asteroid mining, advanced materials) and governance decisions are tied to shareholder interests.

<u>3. Direct Democracy Model: The "Participatory Habitat"</u>

<u>Overview</u>

This model envisions a <u>direct democracy</u>, where every resident has an active role in governance by voting directly on laws, policies, and resource allocation.

<u>Key Features</u>

- <u>Citizen Assemblies</u>: All residents participate in decision-making through regular assemblies or digital platforms.

- Consensus-Oriented Governance: Decisions are made through consensus or majority voting.
- Rotating Leadership: Leaders are selected temporarily from the population to facilitate administrative tasks.
- Digital Platforms: Advanced technology allows for real-time, large-scale citizen participation.

Strengths

- Maximizes citizen engagement and transparency.
- Encourages collective responsibility for the habitat's success.
- Avoids the concentration of power in a single entity.

Weaknesses

- Decision-making can be slow and inefficient, especially in emergencies.
- May encounter challenges in scaling to larger or more diverse populations.
- Risk of "tyranny of the majority" without protections for minority rights.

Example

A small-to-medium-sized habitat designed around collaborative decision-making, with residents voting on everything from resource distribution to cultural programming.

4. Technocratic Model: The "AI-Managed Habitat"

Overview

In this model, governance is primarily handled by artificial intelligence systems designed to optimize resource management, quality of life, and conflict resolution. Human oversight serves as a secondary layer.

Key Features

- <u>AI Leadership</u>: Advanced AI systems regulate the habitat's ecosystem, economy, and infrastructure based on data-driven algorithms.
- <u>Human Oversight Boards</u>: Committees of experts oversee AI decisions and intervene when necessary.
- <u>Efficiency and Stability</u>: AI ensures optimal allocation of resources and eliminates human bias.
- <u>Predictive Governance</u>: AI uses predictive models to anticipate problems and implement solutions proactively.

Strengths

- Highly efficient and impartial decision-making.
- Minimizes human error and corruption.
- Ensures sustainability in closed-system environments.

Weaknesses

- Risk of over-reliance on AI, with potential for catastrophic failure if systems malfunction.
- Limited human agency in governance decisions.
- Ethical concerns about AI control over human lives.

Example

A habitat governed by an AI system that monitors air quality, food production, and population dynamics while consulting with human experts for policy adjustments.

5. Charter Model: The "Founding Compact"

Overview

Each O'Neill habitat operates under a <u>founding charter</u> that establishes its governance structure, rights, and obligations. These charters are designed to reflect the unique goals and values of each habitat.

Key Features

- Diverse Governance Systems: Habitats may choose different governance models (e.g., democracy, technocracy, or corporate governance) based on their founding charter.
- Freedom of Association: Residents can choose to move between habitats to find governance systems that align with their preferences.
- Inter-Habitat Agreements: Habitats collaborate on shared goals (e.g., trade, security) through treaties and alliances.
- Charter Amendments: Governance structures can evolve over time through charter revisions.

Strengths

- Encourages innovation and experimentation in governance.
- Allows for cultural and political diversity across habitats.
- Provides residents with the freedom to "vote with their feet."

Weaknesses

- Lack of centralized coordination could lead to fragmentation.
- Risk of inequality between habitats with differing resources or governance systems.
- Potential for disputes over shared resources or external threats.

Example

A network of habitats with distinct charters: one focused on environmental sustainability and direct democracy, another on technological innovation and corporate governance.

6. Global Earth Oversight Model: The "Earth Mandate"

Overview

O'Neill habitats remain under the jurisdiction of Earth-based governments or international organizations, such as the United Nations, which set policies and oversee governance.

Key Features

- UN or Earth Oversight: Governance frameworks are established by Earth organizations to ensure accountability and fairness.
- Habitat Administrators: Each habitat is managed by Earth-appointed administrators or councils.
- Universal Standards: Earth-based authorities enforce universal rights, trade regulations, and environmental protections.

Strengths

- Prevents the rise of unregulated or exploitative systems in space.
- Ensures alignment with Earth's values and interests.
- Simplifies conflict resolution through a centralized authority.

Weaknesses

- Risk of Earth-centric governance that disregards the unique needs of space habitats.
- Potential for resistance or independence movements among habitats.
- Bureaucratic inefficiency and delays in decision-making.

Example

A United Nations Space Authority (UNSA) oversees all O'Neill habitats, setting universal laws while allowing local governance for day-to-day operations.

Conclusion

The governance of O'Neill habitats will ultimately depend on their scale, population size, cultural diversity, and economic goals. A hybrid approach combining elements of these models may prove most effective, allowing for autonomy, efficiency, and adaptability. As humanity expands into space, ensuring that these governance systems prioritize fairness, sustainability, and innovation will be critical to building a thriving, extraterrestrial civilization.

PFIT-Federation Governance Model for
O'Neill Extraterrestrial Habitats

(PFIT: Peace and Freedom, and
Intentional Transparent communities)

Introduction

The PFIT model (Peace and Freedom, and Intentional Transparent communities) envisions a governance structure for extraterrestrial societies that balances individual liberty, community diversity, and societal cohesion. This hybrid governance system aligns with the unique requirements of O'Neill habitats—large, self-sustaining, extraterrestrial environments—by emphasizing freedom of movement, community-level autonomy, and societal-level enforcement of peace and transparency. The PFIT-Federation Governance Model combines the core principles of PFIT with federalist and charter-based elements to ensure a fair, adaptable, and scalable system for governing a network of space habitats.

Core Principles of the PFIT-Federation Model

1. Peace and Nonviolence:

- The overarching society enforces a strict ban on weapons, weapons-making, violence, and animal cruelty. This principle ensures that all communities coexist peacefully and that individual rights are protected universally.
- Mechanisms are in place to prevent and resolve conflicts between individuals, communities, and habitats without resorting to violence.

2. Freedom and Mobility:

- Every individual has the freedom to leave any community and join another or establish their own intentional community. This ensures that personal liberty is safeguarded across the society.
- Each person is fully informed of their right to leave, and communities are required to facilitate the peaceful exit of

- 360 -

members, including providing resources for relocation or temporary support.

3. Intentional Communities:

- Communities within the society are voluntary and self-governing. Each community determines its own membership criteria, governance style, cultural norms, and internal laws, provided they adhere to the broader societal principles of peace and freedom.
- Communities are free to experiment with various governance models (e.g., democracy, technocracy, meritocracy) and cultural structures, fostering diversity and innovation.

4. Transparency:

- Both the overarching society and individual communities must operate with full transparency. This includes ensuring that all individuals are aware of their rights and responsibilities, as well as providing mechanisms for fair and open decision-making.
- Communities are required to enforce and communicate the societal principles of peace and freedom in a good-faith, transparent manner.

5. Unlimited Resources and Space:

- The abundance of "free land" (or habitat space) in the extraterrestrial context prevents the territorial and resource conflicts common in Earth's history. This allows for peaceful coexistence and provides individuals the opportunity to create new communities or live independently as "hermit" communities.

The Two-Tier Governance Structure

The PFIT-Federation Model operates on two levels:

1. Society Level (Extraterrestrial Society of Intentional Communities):

- The overarching society enforces universal principles of peace and freedom, ensuring that all communities coexist harmoniously.
- The society is responsible for maintaining inter-community transparency, arbitration of disputes, and the enforcement of nonviolence.

2. Community Level (Intentional Communities):

- Communities have wide latitude to govern themselves according to their own charters, provided they respect societal principles.
- Communities experiment with different cultural, ethical, and governance systems, creating a dynamic and evolving ecosystem of ideas.

I. Society-Level Governance

At the level of the Extraterrestrial Society of Intentional Communities, the focus is on enforcing universal principles while allowing communities autonomy.

Society Responsibilities:

1. Enforcement of Core Principles:

- Ban on weapons, violence, and animal cruelty is universally enforced.
- Ensure that all individuals are fully informed of their right to leave any community.

2. Conflict Resolution and Arbitration:

- The society provides nonviolent mechanisms for resolving disputes between individuals, communities, or habitats.
- A Conflict Arbitration Council mediates disputes fairly and transparently.

3. Resource Allocation and Infrastructure:

- The society facilitates the allocation of common resources (e.g., energy, air, water) and management of shared infrastructure (e.g., inter-habitat transportation).

4. Transparency Mechanisms:

- A Transparency Oversight Committee ensures that all communities are adhering to societal principles and that individuals are informed of their rights.
- Public records and decisions are accessible to all residents.

Society Governance Structure:

- Universal Charter: A founding document outlines the principles of peace, freedom, transparency, and intentionality. All communities must adhere to this universal charter.
- Representative Assembly:
  - Each community elects representatives to a Society Assembly that oversees the enforcement of societal principles.
  - Representation is proportional to population size, ensuring fairness and inclusivity.
- Executive Council:
  - A smaller executive body handles day-to-day operations, such as resource allocation, security, and external diplomacy.
- Judiciary:
  - An independent judiciary ensures adherence to the universal charter and resolves disputes.

II. Community-Level Governance

Each Intentional Community operates autonomously within the framework of the society, allowing for diverse governance models and cultural experimentation.

Community Responsibilities:

1. Membership and Intentionality:

   - Communities define their own membership criteria, values, and governance systems. Members voluntarily join and agree to abide by community rules.
   - Communities must establish mechanisms to inform members of their right to leave and facilitate peaceful exits.

2. Internal Governance:

   - Communities have full autonomy to experiment with governance styles. Examples include:
     - Democratic Communities: All members vote on decisions.
     - Technocratic Communities: Experts are appointed to lead based on merit.
     - Cultural Communities: Governance based on shared traditions or philosophies.
     - Hermit Communities: Single-person communities where individuals govern themselves.

3. Transparency and Good Faith:

   - Communities must operate transparently, ensuring that all members understand their rights and responsibilities.
   - Communities must enforce societal principles of peace and freedom in good faith.

Community Charters:

- Each community operates under a founding charter that defines its governance structure, values, and membership policies.
- Charters are amendable through internal mechanisms such as referenda or council decisions.

III. Mobility and Freedom

Freedom of Movement:

- Individual Liberty: Each individual has the freedom to leave any community at any time, ensuring that no one is trapped in a system they do not consent to.
- Support for Relocation: The society and communities work together to provide resources for individuals transitioning between communities or establishing new ones.

Founding New Communities:

- Individuals are free to establish their own communities, provided they adhere to the societal principles of peace and transparency.
- New communities are granted access to habitat space and resources to ensure their viability.

IV. Conflict Resolution

Inter-Community Disputes:

- The Conflict Arbitration Council mediates disputes between communities to ensure peaceful resolutions.
- Arbitration decisions are transparent and binding, with opportunities for appeal to the judiciary.

Internal Disputes:

- Communities handle their own disputes internally, but individuals retain the right to appeal to the society-level judiciary if societal principles are violated.

Advantages of the PFIT-Federation Model

1. Peaceful Coexistence: Universal enforcement of nonviolence ensures stability and security.
2. Individual Freedom: Individuals retain full autonomy to choose or leave communities.

3. Cultural Diversity: Communities can experiment with unique governance models and cultural practices.
4. Transparency: Both society and communities operate openly, ensuring accountability.
5. Innovation: Communities serve as laboratories for new ideas, fostering progress and adaptability.

Conclusion

The PFIT-Federation Model combines the strengths of peace, freedom, and intentionality with the benefits of transparency and flexibility. By fostering a diverse ecosystem of autonomous communities while ensuring universal principles of nonviolence and liberty, this governance system provides a practical and fair framework for governing O'Neill habitats and extraterrestrial societies. This hybrid model creates a foundation for a peaceful, innovative, and inclusive future in space.

Pathway to the PFIT-Federation World:
Establishing a Strongly Enforced Space Treaty

Transitioning from our current geopolitical and technological reality to a PFIT-Federation world—a society of peaceful, free, and intentional communities in extraterrestrial space—requires a deliberate, phased approach. Central to this transition is the establishment of a Space Treaty that bans weapons and their manufacture in extraterrestrial space, enforced by an independent, non-state entity such as the Agency for a Better Cosmos (ABC). Below is a roadmap for achieving this vision.

1. Building on Existing Space Law Frameworks

The foundation for a weapon-free extraterrestrial space already exists in the form of the Outer Space Treaty (OST) of 1967. This treaty, signed by over 100 nations, prohibits the placement of weapons of mass destruction (WMDs) in outer space and mandates the peaceful use of celestial bodies. However, the OST has limitations:
- It does not explicitly ban conventional weapons in space.

- It lacks strong enforcement mechanisms.
- It primarily governs state actors, leaving private entities and individuals unregulated.

Action Steps:

1. Expand the Outer Space Treaty:

   - Convene an international summit to negotiate an updated treaty or a new complementary treaty that explicitly bans all weapons and their manufacture in extraterrestrial space.
   - Include provisions for regulating private actors and non-state entities, ensuring comprehensive coverage.

2. Incorporate Lessons from the Artemis Accords:

   - The Artemis Accords, signed by multiple nations in 2020, emphasize peaceful exploration and cooperation in space. Use this framework to encourage collaboration among nations and private entities in drafting the new treaty.

3. Leverage the PAROS Initiative:

   - The Prevention of an Arms Race in Outer Space (PAROS) initiative, supported by the United Nations, seeks to prevent the weaponization of space. Build on this initiative to gain international consensus on banning weapons in extraterrestrial space.

2. Establishing the Agency for a Better Cosmos (ABC)

The Agency for a Better Cosmos (ABC) would be an independent, international organization tasked with enforcing the Space Treaty. Unlike traditional enforcement mechanisms tied to nation-states, the ABC would operate as a neutral, non-state entity with the authority to act against violations by any actor—state or non-state.

Key Features of the ABC:

1. Independence:

- The ABC operates independently of any single nation or bloc, ensuring impartial enforcement of the treaty.
- Its funding comes from contributions by signatory states, private space companies, and international organizations.

2. Enforcement Mechanisms:

- The ABC is equipped with advanced monitoring technologies, such as satellite surveillance and AI-driven anomaly detection, to identify violations in real time.
- It has the authority to impose penalties, confiscate illegal equipment, and, if necessary, disable unauthorized weapons systems in space.

3. Global Participation:

- Membership in the ABC is open to all nations, private entities, and international organizations that endorse the Space Treaty.
- Non-signatories are still subject to the treaty's enforcement, ensuring universal applicability.

4. Transparency and Accountability:

- The ABC operates with full transparency, publishing regular reports on its activities and decisions.
- A global oversight council, composed of representatives from signatory states and independent experts, ensures accountability.

3. Rallying Initial Support for the Space Treaty

To gain momentum, the Space Treaty must initially be signed by a critical mass of influential states and peoples, including major spacefaring nations and emerging space powers. These early adopters will set the stage for broader global acceptance.

Steps to Rally Support:

1. Engage Major Spacefaring Nations:

- The United States, China, Russia, the European Union, Japan, and India are key players in space exploration. Their participation is essential for the treaty's legitimacy and enforcement.
- Highlight the mutual benefits of a weapon-free space, such as reduced risk of conflict and enhanced opportunities for peaceful collaboration.

2. Involve Emerging Space Powers:

- Countries like the UAE, South Korea, and Brazil, as well as private space companies (e.g., SpaceX, Blue Origin), should be included in negotiations to ensure diverse representation.

3. Appeal to Global Values:

- Frame the treaty as a legacy for future generations, emphasizing the moral and practical imperative of maintaining peace in space.
- Highlight the treaty's alignment with global goals, such as the United Nations' Sustainable Development Goals (SDGs).

4. Public Advocacy Campaign:

- Launch a global campaign to raise awareness about the treaty and its benefits, engaging citizens, scientists, and civil society organizations.
- Use media, educational programs, and international forums to build public support.

4. Enforcing the Treaty Universally

Once the treaty is signed and the ABC is operational, enforcement must apply to <u>all actors</u>, regardless of whether they have signed the treaty. This universal enforcement ensures that no one can exploit loopholes or evade accountability.

Enforcement Strategies:

1. Monitoring and Surveillance:

   - Deploy a network of satellites and sensors to monitor space activities and detect violations.
   - Use AI and machine learning to analyze data and identify potential threats.

2. Sanctions and Penalties:

   - Impose economic and diplomatic sanctions on violators, including states, corporations, and individuals.
   - Confiscate or disable unauthorized weapons systems in space.

3. Collaboration with Private Sector:

   - Partner with private space companies to ensure compliance with the treaty.
   - Provide incentives, such as tax breaks or grants, for companies that adhere to the treaty's principles.

4. Conflict Resolution:

   - Establish a neutral arbitration body within the ABC to resolve disputes related to treaty violations.

5. Transitioning to the PFIT-Federation World

With the Space Treaty and the ABC in place, the foundation is laid for the development of a PFIT-Federation world. This transition involves fostering the growth of intentional communities in space while ensuring adherence to the principles of peace, freedom, intentionality, and transparency.

Steps for Transition:

1. Encourage the Formation of Intentional Communities:

- Provide resources and support for individuals and groups to establish intentional communities in space.
- Ensure that each community operates transparently and respects the principles of peace and freedom.

2. Promote Technological and Economic Development:

- Invest in technologies that enable sustainable living in space, such as closed-loop life support systems and renewable energy.
- Facilitate inter-community trade and collaboration to build a thriving extraterrestrial economy.

3. Foster a Culture of Peace and Freedom:

- Educate future space settlers about the principles of the PFIT-Federation model.
- Encourage cultural exchange and dialogue between communities to promote mutual understanding and cooperation.

4. Expand the ABC's Role:

- As the number of space communities grows, the ABC evolves into a broader governance body, coordinating inter-community relations and ensuring adherence to PFIT principles.

Conclusion

The journey to a PFIT-Federation world begins with the establishment of a strongly enforced Space Treaty that bans weapons and their manufacture in extraterrestrial space. By rallying support from key states and peoples, creating the independent Agency for a Better Cosmos (ABC), and fostering the growth of intentional communities, humanity can lay the groundwork for a peaceful, free, and transparent extraterrestrial society. This vision represents a profound legacy for future generations, ensuring that space becomes a realm of cooperation, innovation, and harmony.

Promoting an Optimal AI and Media Environment to Achieve the Space Treaty and PFIT-Federation Goals

To achieve the ambitious goals of a Space Treaty banning weapons in extraterrestrial space and the establishment of a PFIT-Federation world (Peace and Freedom, and Intentional Transparent communities), it is essential to create an environment where artificial intelligence (AI) and media systems are leveraged effectively. These tools can play a transformative role in shaping public opinion, fostering international collaboration, and ensuring transparency and accountability. Below is a roadmap for promoting an optimal AI and media environment to support these goals.

1. Leveraging AI for Advocacy and Governance

AI has the potential to accelerate progress toward global goals, including the establishment of peaceful and cooperative extraterrestrial societies. However, its use must be carefully managed to avoid risks such as bias, misinformation, and misuse.

1.1. AI for Public Awareness and Advocacy

- AI-Driven Campaigns: Use AI to design and optimize global advocacy campaigns for the Space Treaty. AI can analyze public sentiment, identify key influencers, and tailor messaging to resonate with diverse audiences across cultures and languages.
- Personalized Education: AI-powered platforms can educate individuals about the benefits of a weapon-free extraterrestrial space and the principles of the PFIT-Federation. For example, interactive AI tools can simulate life in intentional communities, helping people understand the vision and its practical implications.

1.2. AI for Transparency and Monitoring

- Global Monitoring Systems: AI can be used to monitor space activities in real time, ensuring compliance with the Space Treaty. For example, machine learning algorithms can analyze satellite data to detect unauthorized weapons manufacturing or deployment.
- Data Transparency: AI systems can process and present complex data about space governance in an accessible and transparent way, fostering trust among stakeholders.

1.3. AI for Collaborative Governance

- AI-Driven Policy Simulations: AI can model the potential outcomes of different governance structures for the PFIT-Federation, helping policymakers design systems that maximize peace, freedom, and sustainability.
- AI for Conflict Resolution: AI tools can mediate disputes between nations or communities by analyzing historical data and proposing fair, evidence-based solutions.

2. Creating a Media Ecosystem for Global Engagement

Media plays a critical role in shaping public opinion and building the global consensus needed to support the Space Treaty and PFIT-Federation goals. A strategic media ecosystem can amplify the message, counter misinformation, and inspire collective action.

2.1. Global Media Campaigns

- Unified Messaging: Launch a global media campaign emphasizing the moral and practical benefits of a weapon-free extraterrestrial space. Highlight the legacy this would leave for future generations and the opportunities for peaceful collaboration.
- Storytelling and Narratives: Use compelling stories to illustrate the vision of the PFIT-Federation. For example, documentaries, films, and virtual reality experiences can immerse audiences in the possibilities of intentional communities in space.

2.2. Countering Misinformation

- AI-Powered Fact-Checking: Deploy AI systems to identify and counter misinformation about the Space Treaty and PFIT-Federation goals. These systems can flag false narratives and provide accurate, evidence-based information.
- Media Literacy Programs: Educate the public on how to critically evaluate information about space governance, ensuring they are not swayed by propaganda or fearmongering.

2.3. Inclusive Media Platforms

- Global Participation: Create media platforms that allow individuals from diverse backgrounds to contribute their perspectives on space governance. This inclusivity fosters a sense of ownership and shared responsibility for the Space Treaty and PFIT-Federation goals.
- Localized Content: Tailor media content to different cultural contexts, ensuring that the message resonates with audiences worldwide.

3. Building International Collaboration Through AI and Media

The success of the Space Treaty and PFIT-Federation goals depends on broad international collaboration. AI and media can facilitate this by fostering dialogue, building trust, and aligning stakeholders around shared objectives.

3.1. AI for Diplomacy

- AI-Powered Negotiation Tools: Use AI to assist in treaty negotiations by analyzing stakeholder positions, identifying common ground, and proposing compromise solutions.
- Language Translation: AI-driven translation tools can break down language barriers, enabling effective communication between nations and communities.

3.2. Media for Global Dialogue

- International Forums: Use media platforms to host virtual forums where governments, private companies, and civil society can discuss the Space Treaty and PFIT-Federation goals. These forums should be accessible to all, fostering transparency and inclusivity.
- Showcasing Success Stories: Highlight examples of peaceful and cooperative space initiatives to inspire confidence in the feasibility of the Space Treaty and PFIT-Federation.

4. Ethical and Inclusive AI and Media Practices

To ensure that AI and media serve the goals of peace and freedom, their development and deployment must adhere to ethical principles.

4.1. Addressing Bias and Inequality

- Bias Mitigation: Develop AI systems that are free from bias, ensuring that they promote equitable outcomes for all stakeholders.
- Global Access: Ensure that AI and media tools are accessible to all nations and communities, not just those in the Global North.

4.2. Safeguarding Privacy and Security

- Data Protection: Implement robust safeguards to protect the privacy of individuals and communities using AI and media platforms.
- Cybersecurity: Secure AI and media systems against hacking and misuse, particularly in the context of space governance.

4.3. Promoting Transparency

- Open AI Models: Use open-source AI models to ensure transparency and accountability in their development and application.

- Transparent Media Practices: Media organizations should disclose their sources and methodologies, building trust with their audiences.

5. Immediate Actions to Promote the Optimal Environment

5.1. Establish a Global Coalition

- Form a coalition of governments, private companies, and civil society organizations committed to using AI and media to promote the Space Treaty and PFIT-Federation goals.

5.2. Invest in Research and Development

- Fund research into AI applications for space governance and media strategies for global advocacy.

5.3. Launch Pilot Projects

- Test AI-driven monitoring systems and media campaigns in specific contexts, such as promoting peaceful space exploration or raising awareness about the risks of weaponization.

5.4. Engage the Public

- Use AI and media to create interactive experiences that engage the public in discussions about the future of space governance.

Conclusion

By leveraging AI and media effectively, we can create an environment that fosters global collaboration, transparency, and public support for the Space Treaty and PFIT-Federation goals. These tools, when used ethically and inclusively, have the power to inspire a shared vision of a peaceful and cooperative extraterrestrial future.

Appendixes by Charles Tandy, Ph.D.

Appendix O.
The original essay on which this volume is based.

Shall The Future Repair The Past?
Universal Scientific Resurrection Of The Dead

Charles Tandy

The only thing that is good without qualification is the good will.
− Immanuel Kant

We all do better when we all do better.
− Senator Al Franken

Separation of moral enlightenment and scientific enlightenment is dangerous. Reason demands the integrated advancement of morality and science. Accordingly, all reasonable persons (reason-able beings) have a moral imperative to attempt individual and world betterment, both synchronically ("at a point in time") and diachronically ("over time"). Such reasonable individuals, then, have a purpose. Over many generations, one's own generation plays its own imperative role in quest of the highest good. One's own generation has moral solidarity with all generations, past and future. Each generation is responsible for further enlightenment:

- Moral advancement
- Scientific advancement
- Integrated advancement of morality and science in quest of the highest good

Perhaps one helpful way to picture this pursuit is to think in terms of a Beloved Community. I sometimes think of the Beloved Community as a kind of foundation or foothold on our way toward the highest good; at other times, I think of the highest good as the Beloved Community. That is, from our present vantage point, we can envision a society as more or less fulfilling our present dreams. Yet those living in such a society might perceive imperfections and dream dreams beyond our present ability to imagine. Thus, as I speak to my present self and to our present generation, I believe the image of a Beloved Community serves a useful purpose.

Kant viewed his Critical Philosophy (his Critiquing of Reason) as a "Copernican Revolution" (to use his own words). Copernicus had shattered the medieval geocentric worldview of earthly sin surrounded by heavenly perfection, as in Dante's epic poem. (Later, the Darwinian Revolution would further invalidate the medieval universe.) Kant sought an enlightened worldview that would replace the medieval universe and provide a reasonable account of the validity and limitations of science.

The revolution initiated by Kant places each and all reasonable beings in the central role of transforming the universe from bad to better. Such finite autonomous beings are neither dogmatic (certain) nor skeptical (despairing), but critical. The dogmatist keeps both the baby and the bathwater, while the skeptic discards both. The process of critical analysis-synthesis proactively seeks discernment and enlightenment rather than certainty or despair.

A primary maxim of reason (i.e., of a reason-able or autonomous person) is: Think for yourself. And a corollary rule is: Examine reason so as to gain insight into reason. We humans use abstractions (e.g., categories, narratives, and models) when we reason. Successful use of a particular orientation (an integrated set of categories, narratives, and models) may simultaneously provide one with confidence and tempt one into mistaking it for the whole of reality. Aware of this, reason engages in critical reflection and in moral evaluation for practical action. Aware of the limits of finite reasoners, of the indicative-imperative of moral identity, and of the power of foresighted decision-making, reason seeks answers to both scientific and philosophical issues. Such answers are accompanied by truth and error, and by additional questions of varying quality and fruitfulness.

The use or function of reason, according to Alfred North Whitehead, includes: 1) to survive; 2) to live well; and, 3) to live better-and-better. For Kantians, reason engages in a quest for truth, pattern, meaning, goodness, beauty, coherence, unity. In this quest, finite human reason abstracts beyond the empirical perception of ordinary objects to notions like categories, narratives, models, natural laws, cause-effect, moral principles, infinity, living better-and-better, immortality, free-will, good-will, ill-will, my-self, other-selves, gods, angels, the Ground of Being, great-great-great grandmothers never seen by oneself, the past, the present, the future, astrology, flat earth, phlogiston, spontaneous generation, Aristotelian physics, four bodily humours, hollow earth, alchemy, Ptolemaic universe, phrenology, mathematics, music, science, religion, art, all-of-reality, the Beloved Community, and much more.

Scientific reasoning and moral reasoning make use of non-provable assumptions or regulative ideas in order to advance or develop (as toward the Beloved Community). This is necessary for finite reasoners with limited access to reality – beings who are nevertheless unwilling to give up the quest for enlightenment. Reason can question appearances – such as the relative existence or reality of the extended world generally, specific ordinary objects, the so-called broken stick in water, the sun circling the earth, regulative ideas – almost anything, as Descartes pointed out via his evil demon thought experiment. Nevertheless, by supplying possibly temporary or fallible assumptions or models we seem to gain an appearance of progress over time. In such case, we are conscious and questioning (critical, instead of either dogmatic or skeptical) in what we do as finite creatures in journey toward the unknown.

Kant's Critical Philosophy places reason, in the form of autonomous beings, in charge of foresighted action and universal advancement, moral and scientific. According to the Kantian Revolution, such reason-able persons have a physical structure embedded in a societal culture – meaning they are finite, neither altogether free nor altogether un-free. Due to these limitations, such beings are blind to those aspects of reality that do not fit (that are beyond) their physical abilities or personal traits. Nevertheless, these abilities and traits at least allow the possibility of moral and scientific advance over time and the further realization of freedom and autonomy in quest of the Beloved Community.

Thus what we think we know about morality or science at a given point in time will presumptively not be true in an absolute sense, yet nevertheless may lead us toward greater enlightenment. Based on who we are physically and culturally, we finite creatures discover-invent principles of morality and laws of nature. These principles and laws, as perceived by finite physical-cultural beings, may not be exactly or absolutely correct, yet they may lead us toward the discovery-invention of more enlightened quasi-standards (ethical principles and scientific laws).

According to Descartes, "I think, therefore I am." What he meant was: I am aware that I am feeling and thinking – therefore, **I** **exist**. According to Kant, I am aware that I am morally free – therefore, **we** **exist**. In the 20th century, Albert Camus commemorated this "I" in his book entitled *The Rebel*. Like the Camusian rebel in quest of right, I too rise up against the benign indifference of the universe: I rebel – therefore, we are. Thus, morality requires discovery-invention of the Beloved Community. I ought, I can: We ought, we can. So there is a sense in which the moral imperative requires me to do the impossible: I have a role to play in the historical drama that transforms what is presently not-possible into what ought be possible. One's own generation has a role to play in the historical drama that transforms what is presently not-possible into what ought be possible, the Beloved Community. That this impossible power is possible is an insight or conclusion reached by Kant in his *Critique of the Power of Judgment*.

For Kant, such considerations are tied to his notion of God; the moral imperative demands that (the benign indifference of) the universe is made or structured so as to both physically permit and morally require – over many generations – that autonomous beings discover-invent the Beloved Community. A perfect being that is infinitely good, that always has been infinitely good, that always will be infinitely good, and that has no alternative but to be infinitely good – may be awe-inspiring, holy, beyond our comprehension. A finite reasonable person may not be holy, but may nevertheless engage in a quest to discover-invent the Beloved Community. If somehow such a finite being were to become infinite and perfect, would this not be awe-inspiring? Anyway, would not the best of all possible universes have to include at least some beings structured so as to be free to make moral and immoral judgments?

According to Kierkegaard, there are many ways to make wrong judgments; "one can be deceived in believing what is untrue, but on the other hand, one is also deceived in not believing what is true." [K, p.23] "If it were true – as conceited shrewdness, proud of not being deceived, thinks – that one should believe nothing which he cannot see by means of his physical eyes, then first and foremost one ought to give up believing in love. If one did this and did it out of fear of being deceived, would not one then be deceived?" [K, p.23] "To cheat oneself out of love is the most terrible deception; it is an eternal loss for which there is no reparation." [K, p.23] Too, Kierkegaard once suggested we take care that the great indicative-imperative path, the beloved way of life, "is not misrepresented by making either the difficulty or the ease too great." [K, p.19] Kant has sometimes been accused of making the practice of "I ought, therefore I can" incredibly, even impossibly, difficult. My account above indicates, I think, a moral path of rational hope that is neither easier nor harder than it ought to be. We are finite and imperfect in our quest for the Beloved Community, yet we have a vital role to play in the great adventure.

Both Thomas Kuhn and Karl Popper pointed out that science is "as if" truth. All past and existing evidence may seem to confirm our scientific theory. Yet future evidence – and/or a new interpretation of old evidence – may seemingly disconfirm the established theory. That is, we are always revising our accounts of the past and of the future. We are always "changing" the past! Take, for example, the following past event: I see a man pulling out a gun and shooting it at a woman; soon thereafter, I see someone saying the woman is dead. How might this past event be revised or changed? (1) With respect to the event, the victim is alive; the victim was reported dead so as to satisfy the emotional needs of the shooter. (2) With respect to the event, the shooter is not a man but is a woman disguised as a man. (3) With respect to the event, I discover that I have come across a theater crew making a movie. Death was only in the movie. (4) With respect to the event, I discover that I have been dreaming. The dream was real only as a dream. (5) With respect to the event, I discover that sometime after the victim was declared dead, they saw her finger spontaneously move; then they detected a heartbeat. She was clinically dead but is now alive.

It seems that time is "reversible" or that the past is "redeemable" or repairable – if we choose to look at events as "full, fat, opaque." [L, p.149] According to Nathaniel Lawrence: "The general pattern is thus quite clear. The more predominant the element of consciousness [reason], the more discontinuous it is as compared to the continuous functioning of the physical elements to which it belongs. The element of consciousness is also less predictable; its activities are less inclined to be law-like, even in pathological cases. It is more free. ... [For Socrates,] the unexamined life [i.e., life without reason] was not worth living." [L, p.163] As I seek to apply such insights or conclusions toward realization of the Beloved Community, I again quote Nathaniel Lawrence: "Actively purposive consciousness [the examined life of reason], concerned with the realization of value, proceeds by ever larger temporal blocks." [L, p.165] "The reversibility of time lies in this, that **what is over and done with, in its relation to some particular temporal process, may quite easily be as yet incomplete with respect to some supervenient process** whose unit blocks are temporally longer, and for which it may have significance still. The past is irrevocable only if the perspective on it be arbitrarily confined." [L, p.166, emphasis in the original]

Thus, given the death of a loved one, we do **not** seek closure. On the contrary, we make use of the power of judgment – we proceed morally toward scientific resurrection of the dead in the context of proactive realization of the Beloved Community. This requires scientific advance in both the physical and social sciences. Our well-intentioned actions may yield crisis/bad results if we lack empirical interactive knowledge of how the complex system works globally or cosmically. We cannot realize the Beloved Community without such necessary hard work of morally-directed scientific-technological advance. The sooner we begin work on the long-term project, the less long-term the project will be.

Brief Reference List

C = Cortese, Franco, "Kant's Theory Of Morality Necessitates Personal Immortality" (chapter 4, pages 87-94) IN: *Death And Anti-Death, Volume 14*. Charles Tandy, editor. (2016) Ann Arbor, MI: Ria University Press. Also available at <https://ssrn.com/abstract=3113867>.

K = Kierkegaard, Soren. *Works of Love*. Howard V. Hong and Edna H. Hong, editors and translators. (1847, original Danish – 1964, English translation) Harper Torchbooks, paperback.

L = Lawrence, Nathaniel. "Time, Value, and the Self" (chapter 5, pages 145-166) IN: *The Relevance of Whitehead*. Ivor Leclerc, editor. (1961) London: George Allen & Unwin Ltd.

T = Tandy, Charles [editor]. [ChatGPT 4o mini –and– Charles Tandy] 2024. "Retroactive Continuity And Universal Resurrection: Philosophical And Technological Pathways" ~~(a chapter in the present volume)~~ IN: *Death And Anti-Death, Volume 22: In Honor Of Saul Kent (First Life Cycle 1939-2023)*. Charles Tandy, editor. ~~Also available at <https://papers.ssrn.com/sol3/papers.cfm?abstract_id=5074924>.~~ [Ann Arbor, MI: Ria University Press.]

ORCID

Charles Tandy <https://orcid.org/0000-0003-2174-6082>.

Shall The Future Repair The Past?
Universal Scientific Resurrection Of The Dead

Charles Tandy

Tandy, Charles. 2024. "Shall The Future Repair The Past?
Universal Scientific Resurrection Of The Dead" is a chapter
contribution in: *Death And Anti-Death, Volume 22: In Honor Of
Saul Kent (First Life Cycle 1939-2023)*. Charles Tandy, Editor.
Ria University Press, Ann Arbor, MI.

Appendix P.
Bibliography for this volume.

This bibliography is hugely deficient or incomplete – i.e., I have listed some authors/books that come to mind. No doubt there are books I might or should have listed below that are not listed because I have not read them or have found myself unable to understand them – not to mention authors and works I am completely unaware of.

In each case below I list only one work by each author. Many (not all) of the authors have written additional books also relevant (directly or indirectly) to the Common Task, our duty of universal scientific resurrection. Relevant publications in the form of articles or book chapters are numerous, too numerous to mention in the present context.

Bandyopadhyay, Anirban
Nanobrain: The Making of an Artificial Brain from a Time Crystal

Bostrom, Nick
Superintelligence: Paths, Dangers, Strategies

Boulding, Kenneth E.
The Image: Knowledge in Life and Society

Brown, Norman O.
Life Against Death: The Psychoanalytical Meaning of History

Camus, Albert
The Rebel

Chalmers, David J.
Reality+: Virtual Worlds and the Problems of Philosophy

Choron, Jacques
Death and Western Thought

Cordeiro, José
The Death of Death: The Scientific Possibility of Physical Immortality and its Moral Defense

De Grey, Aubrey
Ending Aging: The Rejuvenation Breakthroughs That Could Reverse Human Aging in Our Lifetime

Dennett, Daniel C.
From Bacteria to Bach and Back: The Evolution of Minds

Deutsch, David
The Fabric of Reality: The Science of Parallel Universes--and Its Implications

Drexler, Eric
Engines of Creation: The Coming Era of Nanotechnology

Ettinger, Robert C. W.
The Prospect of Immortality

Fahy, Gregory M.
The Future of Aging: Pathways to Human Life Extension

FM-2030
Optimism One

Fyodorov, Nikolai [N. F. Fedorov]
[Nikolay Fedorovich Fedorov] [Николай Федорович Федоров]
(See nffedorov.ru)

Gawdat, Mo
Scary Smart: The Future of Artificial Intelligence and How You Can Save Our World

Greene, Brian
The Hidden Reality: Parallel Universes and the Deep Laws of the Cosmos

Hanson, Robin
The Age of Em: Work, Love, and Life When Robots Rule the Earth

Harari, Yuval Noah
Nexus: A Brief History of Information Networks from the Stone Age to AI

Harrington, Alan
The Immortalist: An Approach to the Engineering of Man's Divinity

Heidegger, Martin
Being and Time

Hick, John
The Philosophy of Religion

Kant, Immanuel
The Three Critiques

Kierkegaard, Soren
Works of Love

King, Martin Luther [Jr.]
Strength to Love

Kuhn, Thomas S.
The Structure of Scientific Revolutions

Kurzweil, Ray
The Singularity Is Nearer: When We Merge with AI

Lawrence, Nathaniel
Alfred North Whitehead: A Primer of his Philosophy

Leclerc, Ivor
The Relevance of Whitehead

Li, Jack [Jack Lee]
Can Death Be a Harm to the Person Who Dies?

Lovelock, James
Novacene: The Coming Age of Hyperintelligence

MacIntyre, Alasdair
After Virtue: A Study in Moral Theory

Minsky, Marvin
The Society of Mind

Moravec, Hans
Mind Children: The Future of Robot and Human Intelligence

More, Max
The Transhumanist Reader: Classical and Contemporary Essays on the Science, Technology, and Philosophy of the Human Future

Nagel, Thomas
Mortal Questions

O'Neill, Gerard K.
2081: A Hopeful View of the Human Future

Parfit, Derek
Reasons and Persons

Penrose, Roger
The Road to Reality: A Complete Guide to the Laws of the Universe

Perry, R. Michael
Forever for All: Moral Philosophy, Cryonics, and the Scientific Prospects for Immortality

Pringe, Hernán
Critique of the Quantum Power of Judgment: A Transcendental Foundation of Quantum Objectivity

Rawls, John
A Theory of Justice

Sandberg, Anders
Whole Brain Emulation: A Roadmap

Segall, Paul
Living Longer, Growing Younger

Sîn-lēqi-unninni
The Epic of Gilgamesh

Stendahl, Krister
Immortality and Resurrection

Suleyman, Mustafa
The Coming Wave: Technology, Power, and the Twenty-first Century's Greatest Dilemma

Susskind, Leonard
The Black Hole War: My Battle with Stephen Hawking to Make the World Safe for Quantum Mechanics

Tandy, Charles
Death And Anti-Death, Volume 22: In Honor Of Saul Kent (First Life Cycle 1939-2023)

Tegmark, Max
Life 3.0: Being Human in the Age of Artificial Intelligence

Teilhard de Chardin, Pierre
The Phenomenon of Man

Tillich, Paul
The Courage to Be

Tipler, Frank J.
The Physics of Immortality: Modern Cosmology, God, and the Resurrection of the Dead

Unamuno, Miguel de
The Tragic Sense of Life in Men and Nations

Whitehead, Alfred North
Process and Reality

Yourgrau, Palle
A World Without Time: The Forgotten Legacy of Godel and Einstein

www.ingramcontent.com/pod-product-compliance
Lightning Source LLC
Chambersburg PA
CBHW022051210326
41519CB00054B/302